Pioneers of the U.S. Automobile Industry

Michael J. Kollins

Other SAE books on this topic:

Ford: The Dust and The Glory, Volumes 1 and 2
by Leo Levine
Order No. R-292.SET

The E-M-F Company
by Anthony J. Yanik
Order No. R-286

The Franklin Automobile Company
by Sinclair Powell
Order No. R-208

For more information or to order this book, contact SAE at
400 Commonwealth Drive, Warrendale, PA 15096-0001
(724) 776-4970; fax (724) 776-0790
e-mail: publications@sae.org
web site: www.sae.org/BOOKSTORE

Pioneers of the U.S. Automobile Industry

Volume 3: The Financial Wizards

Michael J. Kollins

Society of Automotive Engineers, Inc.
Warrendale, Pa.

Library of Congress Cataloging-in-Publication Data

Kollins, Michael J.
    Pioneers of the U.S. Automobile Industry / Michael J. Kollins
      p. cm.
    Includes bibliographical references and index.
    Contents: v. 1. Big three -- v. 2. The small independents -- v. 3. The financial wizards --
v. 4. Design innovators.
    ISBN 0-7680-0904-9 (set) -- ISBN 0-7680-0900-6 (v.1) -- ISBN 0-7680-0901-4 (v.2) --
ISBN 0-7680-0902-2 (v.3) -- ISBN 0-7680-0903-0 (v.4)
    1. Automobile engineers--Biography. 2. Automobile industry and trade--History. 3.
Automobiles--History. I. Title

TL139 .K65 2001
338.7'6292'0922--dc21
[B]

        2001049584

Editing, design, and production by Peggy Holleran Greb
        301 Brookside Road
        Baden, PA 15005

Copyright © 2002 Society of Automotive Engineers, Inc.
        400 Commonwealth Drive
        Warrendale, PA 15096-0001 U.S.A.
        Phone: (724) 776-4841
        Fax: (724) 776-5760
        E-mail: publications@sae.org
        http://www.sae.org

ISBN 0-7680-0902-2 (Volume 3)
ISBN 0-7680-0904-9 (set)

SAE Order No. R-251

Preface

A hundred years ago a trip by automobile was as much a test of manpower as of horsepower; "Men had to be men." In those days, "Get out and get under" (the song was not composed then) had a direct meaning to the adventuresome, soiled, and grease-stained motorist chauffeurs. Happily, these crude, cumbersome, horseless carriages are no more. Here and there a restored one may be found, hidden among the array of glistening new vehicles of modern achievement. Those pioneer vehicles were in fact as naive as the ancient chariots of Egypt and Rome.

The early horseless carriages were big, heavy, uncomfortable, noisy, and smelly wagons or carriages, powered by engines having huge cylinders that were gluttons for fuel. Or they were small, fragile, uncomfortable, noisy, and smelly buggies, powered by small engines, hardly big enough to propel the buggy. Clanking chains rotated the rear wheels, while noisy engines dripped oil like a sieve. They emitted billowing clouds of smoke, as the drab vehicles trembled on wobbly wheels that seemed ready to collapse.

Although the average asking price for one of these "headaches on wheels" was $1,000 or more, the greatest expense came later for maintenance and repairs. These vehicles were plagued with engine, clutch, transmission, steering, brake, wheel, and fuel troubles, let alone problems from the weather. The cost of broken and worn-out parts greatly exceeded the cost of operation. Axle shafts fractured, universal joints failed, crankshafts broke or scored, pistons cracked, cylinders scored or wore rapidly, connecting rods broke, bearings burned out, clutches slipped, transmission gears stripped and chattered. Adjustments and overhaul procedures were common operating procedure. The cost of replacement parts was high because of the lack of standardization and volume.

These strange and crude-looking vehicles spit, coughed, belched, groaned, backfired, and stalled unexpectedly. They were the source of distrust, despair, doubt, ridicule, and embarrassment to their owners. It was soon quite obvious that the smelly, noisy, imperfect, and expensive automobile needed much refinement and performance proof to convince a skeptical public that it was a viable alternative to travel by horse. Fortunately, the novelty, rarity, or scarcity attracted enough buyers to keep some manufacturers in business, while the brilliant minds of these stalwart men worked to solve the problems, and to regain the public's confidence, despite the negatives.

These vehicles were the direct ancestors of our modern-day cars, and from their trials, failures, and successes, came knowledge and improvements. It is the purpose of this publication to acknowledge the accomplishments, give credit to, and honor those various, selfless individuals who risked all their possessions and toiled to acquire a better means of transportation, which has led to a better and fuller life for all Americans.

Specifications

The specifications and other pertinent information in this publication may differ from other contemporary automotive publications. The data and prices shown were taken from official records and/or automobile journals of the time.

The prices as shown are based on the lowest-priced runabout as listed in the ALAM (Association of Automobile Manufacturers) through 1911. From 1912 through 1924 the price shown is for the five-passenger touring car as listed by the NACC (National Chamber of Commerce). From 1925 on, the prices listed were the advertised price for the five-passenger sedan model.

Prior to 1911 the horsepower ratings listed were the ALAM rating, calculated by squaring the engine bore size in inches, multiplying by the number of cylinders, and dividing by 2.5. (The stroke was not considered.) From 1912 through 1924 this responsibility was taken over by the NACC, and the same calculation was used. Beginning in 1925, the horsepower rating was taken over by SAE. However, most manufacturers preferred to use the dynamometer brake horsepower in their publications, referred to in this book as the advertised (adv) horsepower.

The sales volume amounts were taken from the official R.L. Polk Company published records.

Acknowledgments

The author acknowledges the help, cooperation, friendship, encouragement, inspiration, and patience of those who helped to acquire and accumulate the research documents, photographs, and all related information. First of all, I want to thank my wife, Julia A. Kollins, for the patience and encouragement of such an enormous achievement. Also, I thank our children: Michael L. Kollins, Richard P. Kollins, and Laura G. Kollins, for their patience and doing without during this project. Thanks to our grandchildren, namely Monika Kollins, Chloe Beardman, and Fiona Beardman, who gave the inspiration to push ahead.

The author expresses his gratitude for the patience, friendship, teaching, and guidance of his mentors. Among these are his friends, classmates, teachers, professors, journeymen leaders, and the chief engineers, presidents, department directors, and other executives of the automobile manufacturers. These individuals made it possible for the author to obtain the hands-on experience and knowledge of automobile design, engineering, manufacture, and merchandising of automotive products.

Next, the author expresses his thanks to the many libraries, archives, museums, historical societies, and the personal collectors who were willing to share their historical material, data, and related documents. A special thanks to the directors of historical organizations, their curators and librarians, who took a sincere personal interest in this publication and unselfishly gave their assistance.

Due to the enormous size of this publication, it is humanly impossible to include all persons engaged in the automotive industry. However, a sincere effort is made to include the information pertinent to the history of its founding, development, and growth. The author sincerely apologizes to those who may have been omitted.

Individuals

Roy Abernathy
Paul Ackerman
Rodney F. Ackerman
Fred W. Adams
Alex Adastik
George Adastik
Terrence Adderle
Chester Advent
Freddie Agabashian
Mary Alexander
Robert H. Aller
David Andrea
Charlotte Andres
Emil Andres
Mary Atkinson

Lawrence Ball
Jeremy Ball
John Ball

Thomas Ball
Clay Ballinger
John Barnes
Paul Barrett
Oliver E. Barthel
Chloe Beardman
Fiona Beardman
Harold Beardsley
Maurice Beardsley
Mary Benda
John Benedict
Mary Lou Bennethum
William Bennethum
J. Gordon Betz
Albert Bizer
William Bizer
Albert Bloemker
Dushon Boich
Henry Scripps Booth

Max Booth
William Boyd
Albert Brandt
Eric Braun
Helen Braun
Melbourne Brindle
Frank Brisko
Dorothy Broomhall
Gregg Buttermore

Eugene (Jep) Cadou
Dorothy Caldwell
Eula Caldwell
Russell Caldwell
Thomas Cahill
Edna Callahan
Thomas Carnegie
Edna Carpenter
Jack Carson

Russell Catlin
Eugene Cattabiani
James Chaskel
George T. Christopher
Harold Churchill
James Circosta
Ralph Circosta
Rose Circosta
Harley Citron
Joseph J. Cole
LeRoi Cole
George Collick
John A. Conde
William C. Conley
Goerge Connor
Elizabeth Cook
Donald Cummins
Lyle Cummins
Edward P.J. Cunningham
Michael Cunningham
George T. Cutler
Ronald Cutler
James Cypher

Arthur Dau
Donald Davidson
Michael W.R. Davis
Roy Dean
Jack Decker
Raymond Denges
Peter DePaolo
Thomas P. DePaolo
Robert Derleth
Frank DeRoy
John Dovaras
Denny Duesenberg
Fritz Duesenberg
Gertrude Duesenberg
Harlan Duesenberg
Len Duncan
Zora Arkus Duntov
Reeves Dutton
Ralph Dunwoodie
Nicholas Dyken

Helen J. Early
Christopher Economaki
John Eichorn
Donald Eller

Charlene Ellis
Frances Emmons
Erik Erikson
Harold Estelle
William Evans Sr.
Donald Everett

Matthew Fairlie
Harlan Fengler
J. Robert Ferguson
Hugh Ferry
William Finley
Milton Forester
Arthur W. Fowler
Anton J. Foyt
Roy H. Frailing
Kay Frazier
Howard P. Freers
Norman Froelich
Barbara Fronczak
John A. Fugate

George Galster
James (Radio) Gardner
Jerry E. Gebby
Joseph P. Geschelin
Howard Gilbert
Jack Gilmore
Jeffery Godshell
Reneé (Chevrolet) Goeke
David R. Graham
Mitzi Graham
Ronald Grantz
William H. Graves
Karl M. Greiner
Mrs. C. Grimes
Louis (Jerry) Grobe

Roscoe Hadley
Theodore Halibrand
Ira Hall
Samuel Hanks
Robert S. Harper
John J. Harris
Jack Harrison
Harry Hartz
Leslie R. Henry
Franklin Hershey
Jack Hinkle

Hugh W. Hitchcock
Wilma (Ramsey) Hoffman
Gertrude M. Hope
William Hope
Lindsey Hopkins
Raymond House
Ronald Householder
Herbert Howarth
John Hranko
Howard Hruska
Wallece S. Huffman
Frederick Hunty

Helen Cole Imbs
Ralph H. Isbrandt

J.A. Jacks
Frank Jenkins
Marvin Jenkins
Louis Jilbert
Mary Johnson
Myrtle Johnson
Raymond Johnson
Patricia Jones
Charles Jordan
Roam Jordan
Steve Jordan

August Kapecky
Joseph Karschner
John Kay
Ralph Kegler
Kaufman T. Keller
Nancy Kennedy
Mildred Kernen
Russell S. Kettlewell
Lester Kimbrell
Richard Kirchner
Ward Kirkman
James Kirth
Manuel Klasky
Nathan Klasky
Albert Kline
Julia A. Kollins
Laura G. Kollins
Michael L. Kollins
Monika Kollins
Richard P. Kollins
Lee Kollins
Ada Knight

Theodore Lake
Robert Lakin
James Lamb
H.G. Langston
Ben Lawrence
Myron Lawyer
Robert Laycock
P.R. Leatherwood
Harry LeDuc
Matthew Lee
Donald Leeming
Shirley Leeming
Robert H. Lees
Lugjie Lesovsky
David Lewis
Robert A. Lindau
Jim Locke
Joseph D. Lovely
George Lucas
Helen Lucas
Steve Lyman

Claude T. McClure
J. Kelsey McClure
Alvan Macauley
Edward R. Macauly
Raymond Macht
Casmere Maday
Joseph Madek
Freddie Mangold
John Mangold
Jack L. Martin
William Marvel
John Matera
James Melton
Anthony Merchell
Lawrence Merchell
John C. Merchanthouse
Marshall Merkes
William Meyer
Louis Meyer Sr.
Louis Meyer Jr.
Arthur Meyers
Walter Meyers
Jack Miller
Milford Miller
John W. Mills
Thomas Milton Sr.
Thomas W. Milton III

Herbert L. Misch
Guy Monahan
George Moore
Wilma (Ramsey) Musselwhite

Dennis (Duke) Nalon
Adelia Newhouse
Frank Newman
Byron Nichols
John F. Nichols
John Notte III
Robert G. Nowicke

William O'Brien
Rebecca Obsniuk
George J. Oeftger
Mary Lou Oles
William Opdyke
Patrick O'Riley
Cletus O'Rourke
Edward O'Rourke
Carol Osborne
Bruce Owens

Andrew Pardovich
Johnny Parsons
Clyde R. Paton
Eleanor Paton
Mark Patrick
U.E. (Pat) Patrick
Manfred Paul
Johnny Pawl
Curtis Payntor
Howard Pemberton
Marcel Periat
Charles Phaneuf
Frank Plovick
Thomas Poirer
Carl Preuhs
Vera Prindle

Earl Ramsey
William Ramsey
Louis Rassey
Philip Renick
John J. Riccardo
Maurice Rice
John Rinehart
Rita Rittenhouse

Frank Rose
Ross Roy
Frank Russell
August Russo
Robert Russo
Johnny Rutherford

Harry St. Bernard
Aaron Sawyers
Nell Sawyers
Richard Scharchburg
Edward Schipper
Harold Schram
Carmen Schroeder
Gordon Schroeder
Paul Scupholm
Johnny Shannon
William (Wilber) Shaw
Lawrence Shinoda
John Shipman
Igor Sikorsky
Albert Silver
Powell Sinclair
Michael Sizemore
Joseph Slomski
Ruth Smiley
Earl H. Smith
Jack Smith
John Martin Smith
Mike Smith
Peggy Smith
William J. Smyth
William Snively
John Snowberger
Russell Snowberger
Arthur Sparks
William D. Sparks
Tony Spina
Thomas Spragle
Stanley Steiner
Myron Stevens
Peggy Stevens
Raymond Stevens
Steven Stevens
Russell C. Stone
Frank E. Storey
Robert A. Stranahan
Gary M. Stroh
John W. Stroh

Russell Sutton
Peggy Swalls
Robert P. Swanson
Theodore Swiontek

William F. Taylor
Richard Teague
Robert Tebelman
Theodore P. Thomas
David Thompson
Richard Thompson
Thomas Thompson
Donald Thornber
Milton Tibbetts
Michele Tinson
Alexander Tremulis
Floyd Trevis
Robert Trobek
Roy Trowbridge

Bobby Unser
Roy Utley

Charles Van Acker
Eleanor Van Dyne
Leland Van Dyne
Leonard Van Dyne
Andrew Vesely
Joseph Vichich
Patrick Vidan
Robert Vieth
Jesse G. Vincent
Anthony Viviano
Donald Vnasdale
Rolla Volstedt
Axel Von Bergen
Eric A. Von Braun
Magnus Von Braun

Charles R. Wade
Richard Wallen
Roger Ward
Marsden Ware
A.J. Watson
Ronald Watson

I. Milton Watzman
A.W. Webb
Reuben Webb
Erwin A. Weiss
Emerson R. White
Roger H. White
R.L. Williams
Lester Wood
James Wren
Lester Wright

Anthony Yanick
Wilson A. York
Fred M. Young
Louis Young
Bruce Youngs
Smokey Yunick

Steven Zdanski
Susan Zemmens
Richard J. Zimmer
John S. Zink
Raymond Zink

Publications Used in Research

Automobiles of America by Automobile Manufacturers Incorporated, 1962
American Cars, 1805-1942
American Machinist, November 7, 1895
Automobile, Volume I, September 1899, through Volume II, September 1901
The Automobile, Volume VIII, January 1903, through Volume XXXVII, July 1917
Automobile & Automotive Industries, Volume XXXVII, July 1917, through Volume XXXVII, October 25, 1917
Automotive Engineering, SAE, 1952 through 1999
Automotive Industries, Volume XXXVII, November 1, 1917, through Volume LXXVII, December 1942
Automobile & Motor Review, June 1902
Automobile Topics, Volume I #1 through Volume CCLIV, December 1942
Automobile Trade Journal
Automotive News, 1926 through 1999
Cars With Personalities, by John A. Conde, October 1, 1982
Carriage Monthly, 1890 through 1902
Chilton Catalogue and Directory
Chronicle of the Automobile Industry, Automobile Manufacturers Association, 1893 through 1949
Commercial Car Journal, 1948
The Complete Motorist, by Elwood Haynes, 1913-1914
Cycle & Car, 1895 through 1905
Detroit Free Press, 1928 through 1970
Detroit News, 1928 through 1999
Detroit Saturday Night, 1946-1947

Detroit Times, 1928 through 1950
Duesenberg—Mightiest American Motor Car, by J.L. Elbert, 1950
Horseless Age, Volume I #1, January 4, 1902, through Volume XLIV #3, May 1, 1918
Motor Magazine, 1903 through 1926
Motor Age, Volume I #1, October 4, 1900, through Volume CI, July 1940
Motor World Wholesale
R.L. Polk Company, 1921 through 1972
SAE Journal, 1952 through 1999
SAE Transactions, 1952 through 1999
SAE Roster, 1952 through 1999
Scientific American, 1893 through 1999
Smithsonian Institution Bulletin #198
The Story of Speedway, Indiana, Speedway Civic Committee, 1976
History of the Studebaker Corporation, by A.R. Erskine, 1852-1923
Ward's Automotive Quarterly, 1965
Ward's Auto World, 1965 through 1999

Archives, Libraries, and Museums Visited by Personal Contact or Correspondence

Ann Arbor Public Library, Ann Arbor, Michigan
Auburn-Cord-Duesenberg Museum, Auburn, Indiana
Bendix Historical Archives, South Bend, Indiana
Beloit Public Library, Beloit, Wisconsin
Bloomfield Hills Library, Bloomfield Hills, Michigan
Bridgeport Public Library, Bridgeport, Connecticut
Buick Motor Division Archives, Flint, Michigan
Cadillac Motor Car Division Museum, Detroit, Michigan
Carnegie Library of Pittsburgh, Pittsburgh, Pennsylvania
Case Western Reserve Institute Archives, Cleveland, Ohio
Chrysler Corporation Museum, Auburn Hills, Michigan
Children's Museum, Indianapolis, Indiana
Cincinnati Public Library, Cincinnati, Ohio
Cleveland Public Library, Cleveland, Ohio
Cornell University Library, Ithaca, New York
Crawford Automotive and Aircraft Museum, Cleveland, Ohio
Detroit Historical Museum Archives, Detroit, Michigan
Detroit Public Library, Detroit, Michigan, including Burton Biographical Collection, Legal and Patents Collection, and National Automotive Historical Collection
Elkhart Public Library, Elkhart, Indiana
Flint Public Library, Flint, Michigan
General Motors Institute Foundation Archives, including W.C. Durant Papers, Kettering University Archives, C.S. Mott Papers
Henry Ford Museum Archives, Dearborn, Michigan
Freeport Public Library, Freeport, Illinois
Goshen Public Library, Goshen, Indiana
Grosse Pointe Historical Society, Grosse Pointe, Michigan
Hartford Public Library, Hartford, Connecticut
Elwood Haynes Historical Museum, Kokomo, Indiana

Indianapolis Motor Speedway Museum, Indianapolis, Indiana
Kalamazoo Public Library, Kalamazoo, Michigan
Kenosha Public Library, Kenosha, Wisconsin
Milwaukee Public Library, Milwaukee, Wisconsin
New Haven Public Library, New Haven, Connecticut
New Jersey Historical Society, Newark, New Jersey
Newman and Altman Museum, South Bend, Indiana
Novi Motorsports Hall of Fame Museum, Novi, Michigan
Ohio State Library, Columbus, Ohio
R.E. Olds Museum, Lansing, Michigan
Packard Motor Car Company Engineering Records, Detroit, Michigan
Packard Museum, Warren, Ohio
Racine Public Library, Racine, Wisconsin
Rockford Public Library, Rockford, Illinois
Rose-Hulman Institute, Terre Haute, Indiana
South Bend Public Library, South Bend, Indiana
Studebaker Historical Museum, South Bend, Indiana
Syracuse Public Library, Syracuse, New York
University of Toledo Library, Toledo, Ohio
Vigo County Library, Terre Haute, Indiana
Warren-Trumbull County Library, Warren, Ohio
Willard Library, Battle Creek, Michigan
Youngstown Packard Museum, Youngstown, Ohio

Contents

Introduction 1

Thomas White 3

Albert Pope 19

Henry Leland 35

John Willys 51

Benjamin Briscoe 87

Charles Matheson 95

Allison/Fisher/Newby/Wheeler and the Indianapolis Motor Speedway 101

David Parry 115

Hugh Chalmers 125

Harry Jewett 133

Frederick Chandler 141

Edward Rickenbacker 151

E.L. Cord 159

Index 191

About the Author 199

Contents of Other Three Volumes

Volume 1 – The Big Three

Louis Chevrolet
Walter Chrysler
Dodge Brothers
William Durant
Henry Ford
General Motors Corporation

Volume 2 – The Small Independents

Chapin/Coffin/Bezner/Jackson/Hudson/McAneeny
    and the Hudson Motor Car Company
Louis Clarke
Fred and August Duesenberg
Charles and Frank Duryea
Walter Flanders
Harry Ford
Russell Gardner
Graham Brothers
Hupp/Drake/Hastings/Young and the Hupp Motor
    Car Corporation
Kissel Brothers
Joseph Moon
Charles Nash
Ransom Olds
Packard/Joy/Macauley and the Packard Motor Car
    Company
Peerless
George Pierce and Charles Clifton
The Pratt Family and the Elcar Motor Car Company
Studebaker
Harry Stutz
Edwin Thomas

Volume 4 – The Design Innovators

Elmer and Edgar Apperson
Vincent Bendix
James Scripps Booth
Alanson Brush
David Buick
Joseph Cole
Clyde Coleman
Claude Cox
Herbert Franklin and John Wilkinson
Elwood Haynes
Frederick Haynes
Thomas Jeffery
Edward Jordan
Charles King
Howard Marmon
Jonathan Maxwell
Percy Owen
Raymond and Ralph Owen
Andrew Riker
Frank Stearns
Thomas J. and Thomas L. Sturtevant
C. Harold Wills
Alexander Winton

Introduction

The purpose of this publication is not so much to furnish statistics and technical information on automobiles, as these can be found in libraries and many technical publications, but to expose the warm, human, compassionate, and romantic relationship of the persons involved and the products of their labors.

The normal, restless nature of man is such that he is never satisfied with things as they are. Therefore, in his quest for perfection, he has brought about evolutionary changes in transportation methods, not by need alone, but by the compassionate relationship between man and machine.

The historical (chronological) data is furnished primarily for the purpose of establishing the background, trend of the times, and the environment in which these stout-hearted men developed their marvelous machines. The evolution of mobility can be traced to the invention of the wheel, several millenia before the days of Caesar, to make the movement of large masses of material possible.

The first recorded appearance of a self-propelled land vehicle occurred in France in 1769, when Nicholas Cugnot, a French army captain, built a self-propelled, steam-powered tractor to move artillery field caissons.

In Redruth, England, in 1784, an assistant to James Watt (and against his objections) built a three-wheeled land vehicle powered by a high-pressure steam engine. In 1801 Richard Trevithick, with the assistance of André Vivian, built a rear-engine steam-powered carriage, capable of carrying twelve passengers at speeds up to 9 mph. By the late 1820s, with improved economy and road conditions, many steam carriages appeared on England's roads, many large enough to carry 20 passengers, capable of speeds up to 15-20 mph.

Steam propulsion started to create interest in the United States in 1805 when Oliver Evans moved his massive dredge "Orukter Amphibolus" through the streets of Philadelphia and down the banks of the Schuylkill River. The steam-powered driving wheels gave way to paddle wheels that propelled the dredge upstream to the dredging location.

Steam tractors had a great fascination for the author, because even at the age of four, he developed an insatiable love for these mechanical marvels. The most exciting days of his life were the days when the steam tractors, belching smoke from their smoke stacks, would pull the thresher and baler down the lane to the grain stacks and the barn on the farm. Steam tractors performed a remarkable service to man, in the agricultural fields as well as the railroads. Stationary steam power plants provided a source of power for lumbering, pumping, electrical generation, and marine propulsion uses.

The extensive use of self-propelled personal vehicles did not occur until after the discovery and refinement of gasoline in 1859, and the invention of the hydrocarbon-fueled internal-combustion engine. Ironically, the internal-combustion engine principle was invented almost a century before James Watt developed his steam engine.

In 1680 Christian Huygens, a Dutch scientist/astronomer, working on a proposal by Father Jean deHautefeville of Versailles, France, for a device to pump water, used gunpowder for the explosive driving force, acting directly to drive a piston. Steam was used only as a cleaning agent to clean the cylinder after the explosion. Gasoline was unknown at that time. Work was continued by Dionysius Papin and Thomas Newcomen. In 1804 Isaac DeRivaz, a Swiss engineer, was able to obtain a driving force using hydrogen, and built an engine accordingly. It had been said that DeRivaz adapted this engine to a vehicle.

In 1860 Etienne Lenoir, a Frenchman, developed the first practical gas-fueled internal-combustion engine. Lenoir used coal lighting gas for fuel, and ignited it with a spark from an induction coil. In 1862 Lenoir built a vehicle using this engine, and drove it on the roads in France.

While many had made attempts to discredit Lenoir, even a German publication, *Zur Frage der Freien Concurrenz im Gasmotorenbaue*, published in 1883, acknowledged the existence of the Lenoir engine and vehicle. In the famous Selden Patent case against Henry Ford, a working model of Lenoir's vehicle was built to prove antecedence, and the case was won by Henry Ford.

During this period in 1865 Siegfried Marcus, an Austrian, invented and constructed a practical benzene-fueled internal-combustion engine using the two-stroke-cycle principle, and installed the engine in his self-propelled vehicle. He wasn't interested in getting the engine or vehicle patented, he just enjoyed inventing and creating. Chronologically his engine and vehicle might have been the first. The development was continued further by DeLamarre DeBoutteville in France, who in 1883 invented what is known as a carburetor.

Europe had a head start on the United States because many of their hard-surface roads were built centuries before for military purposes. However, the extensive use of the personal self-propelled vehicle was not the result of any single invention or individual, but rather the result of the combined efforts of many inventors, in many countries of the world, who were not aware of what other inventors were doing at the time. The author does not take exception to the method of record of the accomplishments of the automotive inventors; however, for the benefit of the reader we offer the facts as they pertain to the inventor.

The first recorded successful American gasoline-engine-powered vehicle was operated on September 21, 1893, in Springfield, Massachusetts, using the design of Charles and Frank Duryea. The breakthrough for personal self-propelled vehicles did not happen until after the "Spindletop" gusher oil well erupted in 1901, and gasoline prices dropped to such a low level that automobile travel was made affordable.

The development and use of the liquid-fuel-engine-powered vehicle has virtually changed every aspect of American life. The family car has become the American way of life, and it has literally remade rural America. This individual and flexible form of transportation has provided Americans with a new-found freedom, a mobility ending isolation in the city as well as on the farm. This movement gave stimulus for the development of the land and resources, and contributed to the vigorous growth of the American economy. Automotive transportation has certainly enriched the lives of Americans, and has provided jobs for one-seventh of the total United States work force.

The most important aspect of this story is that, while its beginning is not too fully described, and each chapter is concluded, it has no ending. Its fullest meaning lies in the promise of greater accomplishments to come, in the unending story of progress that is the epic of America.

1941 White military and civilian trucks. (Source: White Motor Co.)

Thomas White

Thomas H. White founded one of the most-reputable and longest-lived automotive enterprises in U.S. history. With its roots as a sewing machine company, the White Motor Company grew from one of the first steam-powered trucks to part of today's Volvo-GM Heavy Truck Corporation.

* * *

Sewing Machines Start it All

Thomas H. White was born on April 26, 1836, in Phillipston, Massachusetts. While employed at a chair factory in Templeton, Massachusetts, in 1859 White was granted a patent on a small, hand-operated, single-stitch sewing machine that he invented a couple of years earlier. Soon after he established a shop in Templeton and started to manufacture his machine under the brand label "New England." A year or more of experimenting smoothed the way for successful manufacturing and selling operations. At that time White partnered with William L. Grout, who had selling experience and was given the responsibility of sales for the partnership. Grout remained with White until 1860, when the partnership was dissolved. Progress and financial success continued for White in his individual business enterprise, with the machine gaining widespread recognition and increased sales. Profits from the operation were reinvested for product improvement and expansion of manufacturing facilities.

With orders for his machine increasing, White moved into a larger factory in Orange, Massachusetts, in 1863. As production increased and sales territory spread out, White teamed with a new partner to handle the sales responsibility. In 1866 White moved his business operations to Cleveland, Ohio, because of greater future market potential, better transportation accommodations, and the need for a more central location. A factory building on Canal Street in Cleveland provided the new site of operations and manufacturing. Coincident with this move, the business was organized as the White Manufacturing Company, and sales and profits increased steadily.

Starting in 1867 and for ten years afterward, the White Manufacturing Company also built sewing machine heads for the W.G. Wilson Company, a prominent sewing machine distributor nationwide. This operation substantially added to White's profits. On July 7, 1876, the White Manufacturing Company was reorganized as the White Sewing Machine Company. A new model sewing machine, developed and perfected by two company machinist-mechanics, was introduced. Due to nationwide and international acceptance of White sewing machines, an extensive establishment of sewing machine branch dealers and distributors in the United States was formed, and a foreign office was opened in London, England, in 1880.

By 1881 White production reached 2,000 sewing machines per week, as compared to the original production of 25 machines per month. Rapid growth in overseas markets required the establishment of more foreign offices. In 1884 White Sewing Machine Company diversified into other consumer and industrial products including automatic lathes and screw machines, bicycles, and phonographs. Phonographs and bicycles were a lucrative business at that time, and by 1898, White bicycle production reached 10,000 per year; however, a few years later there was a sudden decline in the bicycle business. White was also heavily involved in the production of bicycle pedals and other components for other bicycle manufacturers. White manufacturing facilities were expanded to meet the need.

The Sons Lead the Drive for Vehicles

In 1892, Windsor T. White (Thomas' eldest son) joined the White Sewing Machine Company and started working in the factory. Son Rollin H. White joined the company shortly after his graduation from Cornell University in 1894. Rollin was eagerly interested in horseless vehicles, therefore his father decided to send him to Europe to learn more about horseless carriages. After spending many months in Europe, Rollin returned in 1898 and started work on developing a new type of flash boiler. The success of his new flash boiler led to the plans for building a new steam vehicle. In 1899, under Rollin's supervision, draftsmen made drawings of the first steam car to be built by the White Sewing Machine Company. By 1900 the first White Steamer (a Stanhope model) and three additional models were made in a corner of the White factory, establishing White as one of the pioneers in the automotive field. The first light steam-powered truck built by White was completed in 1900, and delivered in early 1901 to the Denver Dry Goods Company of Denver, Colorado.

1903 White steam car. (Source: "Land Transport, II. Mechanical Road Vehicles," Science Museum, London, 1936)

Son Walter C. White joined the White Sewing Machine Company in 1900, after acquiring a law degree at Cornell University in 1898 and after two years of legal experience at the New York Central Railroad headquarters in New York City.

The automotive business developed by Windsor, Rollin, and Walter received the full attention, blessing and encouragement of their father. Meanwhile, Thomas continued to direct and spend most of his time with the sewing machine end of the business. In 1901 production of White steam vehicles reached three per week. Work was begun on the first heavy-duty, 5-ton, steam-powered truck. This truck was completed and delivered in early 1902.

Starting in 1901, White vehicles were engaged in competitive auto racing. They won their first race in Detroit, Michigan, with Rollin White driving. Four White Steamers competed successfully in the New York to Buffalo Endurance Run, which terminated at Rochester, New York, due to the assassination of President

1904 White vehicles.
(Source: White Motor Co.)

Jay Webb drove Whistling Willy to the world speed record in 1905. (Source: White Motor Co.)

William McKinley on September 6, 1901. For the next five years, White competed successfully in tours and cross-country competitions including the famous Glidden Tours, New York to St. Louis tour, Chicago to St. Paul road race, and many others. White successfully competed in the 650-mile reliability trials held by the Automobile Club of Great Britain and Ireland.

To obtain a more thorough hands-on business experience, Walter was sent to London, England, to establish White European sales headquarters. Walter was successful in getting European approval of White vehicles and other products of White Sewing Machine Company and established many sales outlets. He remained in London until late 1904, when he rejoined the parent company in Cleveland. In the meantime, sales in the United States climbed dramatically, such that in 1903, White was able to manufacture and sell more than 500 vehicles. In order to keep pace with sales, White established factory branches in Detroit, Boston, New York, Denver, San Francisco, and St. Louis.

On March 4, 1905, the only automobile permitted in the President Theodore Roosevelt Inaugural Parade was a White Steam Car, driven by Walter White. On July 4, 1905, the famous White racing car known as "Whistling Willy" lowered the world speed record for the mile to 48.3 seconds at Mt. Morris Park, New York, with Jay Webb driving. During 1905, White built and sold over 1,000 vehicles.

White steamers and White west-coast representatives aid the Army Relief Corps immediately after the San Francisco earthquake in 1906. (Source: White Motor Co.)

The White Company

In 1906 the White Company was organized, separate from the White Sewing Machine Company, to handle the production and sale of White vehicles. Windsor White was elected president and Walter White vice president. The officers of the White Sewing Machine Company remained as before: Thomas White president, Windsor White first vice president, Walter White second vice president, W.W. Chase secretary, and F.M. Sanderson treasurer. During 1906 over 1,500 White vehicles were built and sold. The variety of vehicles included passenger cars, ambulances, mail trucks, fire trucks, and utility service trucks. White Steamers performed heroically on the days following the San Francisco earthquake on April 17, 1906, helping to provide rescue and relief work for the victims. The vehicles produced during 1907 and 1908 were improved versions of the 1906 models. Production for 1907 was about 1,100 vehicles, and about 1,000 vehicles for 1908.

Gasoline Engines Take Over

In June 1909 White introduced its first gasoline-engine-powered car. The car had a four-cylinder, 25.6-hp (ALAM) engine, and was priced at $2,000 to $2,500 according to body style. The gasoline-engine-powered cars were built concurrently with the steam-powered cars.

During 1909 the White Company Limited, an Ontario Canada company, was organized to conduct the Canadian business. Windsor White was named president, Walter White vice president, A.R. Warner secretary, and Morse Fellers treasurer. The production for 1909 was approximately 1,400 vehicles.

The first White 3-ton, gasoline-engine-powered truck was exhibited at the 1910 New York National Auto Show. Production for the 1910 model year was about 2,200 vehicles, almost equally split between steam and gasoline engine power. On July 1, 1911, White added a new six-cylinder gasoline engine to its line of cars, and the steam-powered engines were gradually phased out.

The Focus on Trucks

Because of the excellent reputation of White, the authorities of Fort Wayne, Indiana, who required all other truck manufacturers to post a bond on the quality of their products, excused the White Company. The resolution, as passed by the city officials, offered this explanation: "Because the reputation of the White Company is a sufficient bond for the quality of its product." In 1912, as the result of witnessing competitive tests by the Russian Army, Czar Nicholas of Russia ordered the army to place an order with the White Company for a fleet of ten 3-ton White trucks. This was the largest order placed by a foreign government for American trucks up to that time. White military vehicles were also used in the Chinese campaign. The White production for the year of 1912 was over 2,800 vehicles.

During 1913 the U.S. Post Office Department purchased its first fleet of twenty 3/4-ton White trucks. White busses went into service for the U.S. Park System, first at Yosemite Park and shortly afterward in the service of nearly all the national parks in the country. White military vehicles accompanied U.S. troops in Haiti and Santo Domingo.

1910 White G-A touring car.
(Source: White Motor Co.)

1910 White G-B Landaulet.
(Source: White Motor Co.)

The Father Steps Down

At a board of directors meeting on December 20, 1913, Thomas White stepped down from the presidency, and new officers of the White Company were elected: Windsor White president, Rollin White first vice president, Walter White second vice president, A.R. Warner secretary, and F.M. Sanderson treasurer. Production for 1913 was over 2,900 units.

Thomas remained active with the company until late May 1914, and he passed away on June 22, 1914. His sons Windsor, Walter, and Rollin were in full control of the operations of the White Company. During 1914 White military trucks took part in the military operations at Vera Cruz, Mexico. White busses were put into service in the Andes Mountains of South America, operating at altitudes up to 17,000 ft. France ordered 600 White military trucks for war services, becoming the first contingent of White military trucks serving the Allied Nations of World War I. The White production for the year was 3,600 vehicles.

Late in 1914, Rollin H. White took a leave of absence, presumably with the intent to relax and rest at his relative's plantation in Hawaii.

White Motor Company

On December 23, 1915, the White Motor Company was incorporated in the state of Ohio, with a capitalization of $16,000,000, as the manufacturing organization, and the White Company became a subsidiary to conduct distribution and sales. The officers of the White Motor Company were the same as those of the White Company. At the board of directors meeting of the White Motor Company on April 22, 1916, M.B. Johnson was elected chairman of the board, Windsor White first vice president, E.W. Hulet second vice president, A.R. Warner secretary, and Otto Miller treasurer. During 1916, letters of patent by the Canadian Government were issued to the White Company Limited, a new corporation, authorized to do business in the province of Ontario.

In 1916 General John J. Pershing's troops used White vehicles to penetrate and traverse 400 miles of difficult, rugged, Mexican terrain in pursuit of Pancho Villa. Production in 1916 was 5,852 vehicles.

World War I

Production in 1917 was devoted almost entirely to fill military orders, building approximately 1,000 passenger cars in seven body types, in addition to about 5,800 trucks. In 1918 the manufacture of White passenger cars was discontinued to permit concentration on trucks and busses, especially for the military. The United States Army adopted the 2-ton White as the class A standard of the U.S. Army. Following exhaustive tests, the 3/4-ton White was adopted as the United States Marine Corps standard, indicating the preference of White military vehicles. A total of over 18,000 White trucks, ambulances, scout cars, and staff cars served the United States and the Allied Nations during World War I, operating on many fronts, transporting men, munitions, food, medical supplies, and personnel over all kinds of roads and blasted-out terrain.

During World War I, Walter White was appointed by the Honorable Newton D. Baker, Secretary of War, as chairman of a committee to coordinate Allied troop truck transportation. He conducted extensive scouting,

White scout car in France during WWI. (Source: White Motor Co.)

surveys, and exploration with Allied Nation officials, and was asked to head up the Interallied Transportation Commission to coordinate and administer all barge, rail, and highway movement of troops, equipment, and supplies. For his service he was cited by the British Government, and was awarded the Croix de Guerre and made a Chevalier of the Legion of Honor by the French Government. After World War I, many of the surplus military vehicles were later purchased and used by utility companies and transportation firms. White production was 6,864 vehicles in 1917, and 10,479 vehicles in 1918.

After the war, White Motor Company operations were adjusted to peacetime conditions, and 11,289 vehicles were built in 1919. The 1920 production resulted in 12,415 vehicles, all trucks.

Growth and Expansion of Trucks and Commercial Vehicles

In late 1920 M.B. Johnson, chairman of the board, died. At a board of directors meeting on May 7, 1921, the White Motor Company elected Windsor White as chairman of the board, WalterWhite president, Thomas H. White II (Windsor's son) vice president, George H. Kelly treasurer, and T.R. Dahl secretary. During the next five years, White concentrated on building the finest heavy-duty trucks and commercial vehicles, making evolutionary improvements as they were developed. In 1924 White Motor Securities Corporation was organized to finance deferred time payment sales. In 1925 the White Motor Realty Company was formed to take over a portion of land and buildings owned by the White Company. The officers of the Securities and Realty companies were the same as that of the White Motor Company. The production totals for the years 1920 through 1925 were as follows: 1920 - 12,415 vehicles, 1921 - 6,727 vehicles, 1922 - 8,217 vehicles, 1923 - 10,740 vehicles, 1924 - 9,478 vehicles, and 1925 - 11,933 vehicles.

In 1926 the new White Model 54 six-cylinder-engine-powered bus was exhibited at the American Electric Railway Association's annual convention. At this convention, the Pennsylvania-Ohio Electric Company, operating a fleet of 109 White busses, was awarded the coveted American Electric Railway Association's Award. During 1927 White heavy-duty trucks equipped with transit concrete mixers appeared on the Pacific Coast highways, where they were hailed as a revolutionary development in the construction industry. That same year Walter C. Teagle of New York and Robert W. Woodruff of Atlanta were named members of the White executive committee. In 1928 the first White steam car, a Stanhope model, was placed on permanent exhibition at the Smithsonian Institution.

On September 29, 1929, after a 19-hour fight for his life, White president Walter C. White died as the result of injuries sustained in an automobile accident. Later that year Robert W. Woodruff was elected chairman of the board and president of the White Motor Company, retaining his connection as president of the Coca-Cola Company of America. Other officers elected at the board meeting were: George H. Kelly vice president and treasurer, H.D. Church vice president of engineering, George W. Smith, Jr., vice president of production, Saunders Jones vice president of distribution, and T.R. Dahl secretary. On December 15, 1930, at the board of directors meeting, Ashton G. Bean, Cleveland industrialist

Walter C. White.

and financier, was elected president of White Motor Company, succeeding Robert W. Woodruff who continued as chairman of the board. The production totals of the years 1926 through 1930 were: 1926 – 12,173 vehicles, 1927 – 10,054 vehicles, 1928 – 9,200 vehicles, 1929 – 9,216 vehicles, 1930 – 6,147 vehicles.

The Depression

In 1931 White pioneered the use of Stellite (developed by Elwood Haynes) bimetal valve seats, screwed-in and bonded in place, extending valve life many times over. During 1931 White Company Ltd. established a factory in Montreal, Canada, to build White trucks and busses for the Canadian market. However, the cruel effects of the Depression were felt at White. The heavy-duty truck market dried up as the result of the losses sustained by potential truck buyers. White vehicle sales dropped to 3,620 in 1931, and the operating losses amounted to $3,234,956.

To obtain a greater share of the truck market, an arrangement was made in 1932 with the Indiana Motors Corporation of Marion, Indiana, whereby the White dealer organization would sell Indiana light-duty trucks along with the White heavy-duty models. In spite of these moves, sales did not increase substantially in 1932. With a production of 3,844 vehicles, White sustained a loss of $3,618,000.

In an attempt to have more operating capital, White Motor Company and Studebaker Corporation made plans for a merger in September 1932. Studebaker acquired 95.11% of White Motor Company stock. There were several organizational personnel changes; however, the identity of White remained intact, with A.G. Bean remaining president of White Motor Company. The actual merger was never completed, because it was blocked for a while by a group of minority stockholders, and in the meantime the Studebaker Corporation went into a friendly receivership on March 18, 1933.

In spite of the limitations imposed by the Depression, in 1933 White designed and developed an entirely new type of city bus. The new bus was powered by an under-floor, horizontally-opposed, twelve-cylinder "pancake" engine, designed and built by White. This new arrangement would set the pattern for years to come. White also introduced the new K series of trucks, employing a more favorable distribution of payloads by placing more weight on the front axle. In the meantime White engineers were developing a new scout car for the U.S. Army Ordnance Department. This vehicle later became the standard for the U.S. Army. By the

White municipal bus.
(Source: White Motor Co.)

middle of 1934, the scout car design was established, developed, and tested. In 1934 White Motor Company acquired a large part of the assets of the White Company, and took over the distribution and sales of White products. Sales improved somewhat, with 4,267 vehicles in 1933, and 6,724 vehicles in 1934.

Robert F. Black

In March 1935 the White Motor Company, with additional new capital, was able to reorganize and separate from the Studebaker Corporation. At the board of directors meeting on April 15, 1935, Ashton Bean was elected chairman of the board and Robert F. Black (former president of Brockway Motor Truck Corporation) was named president of White Motor Company. Other officers elected at the board meeting were: T.R. Dahl vice president and secretary, George H. Kelly vice president and treasurer, and George F. Russell vice president of sales. Unfortunately, the White organization as it was in April lasted only three months. On July 27, 1935, Ashton G. Bean died. At the subsequent board of directors meeting, Black was re-elected president, J.H. Bauman was named vice president of sales succeeding Russell, F.M. Bender was elected vice president and general manager, and W.S. Searles was elevated to secretary. Production for 1935 amounted to 6,820 vehicles.

For 1936 White redesigned and restyled all its vehicles, including the new streamlined coaches. The Baltimore and Ohio Railroad took delivery of 20 new ultra-modern White streamlined coaches to use as connector busses. In 1937 White redesigned its twelve-cylinder "pancake" engine, employing a single casting to include both cylinder banks and crankcase as a unit. This arrangement improved its rigidity and facilitated the machining operations. Work was started on the design and development of the new White Super Power engine, which made its debut in 1938. In 1937 White Motor Company put much of its engineering effort into the design and development of the 6-ton 6x6 prime movers for the U.S. Army Ordnance Department. In 1937 F.S. Baster was appointed chief engineer, and production for that year was 12,117 vehicles.

During 1938 White Motor Company acquired the remainder of the White Company assets. All manufacturing and sales activities were now handled by the White Motor Company. Because 1938 was another recession year, sales plunged again, with production at 5,954 vehicles.

In 1939 White developed the famous WhiteHorse delivery vehicle, with the unique features of an air-cooled engine mounted in the rear integrally with the combined transmission and rear axle.

After months of torturous and rigorous testing, White Motor Company executives were convinced that the White scout cars would meet and surpass the U.S. Army's rigid requirements, and 300 White scout cars were delivered to the U.S. Army Ordnance Department in early 1939, as well as the first 6-ton 6x6 Army prime movers. Production for 1939 was 7,602 vehicles.

The White Super Power engines were used to power all White trucks in 1940. Production was concentrated on heavy-duty military vehicles, including tank carriers delivered to the French and British governments. Production for 1940 was 10,116 vehicles. On September 25, 1940, F.T. Macrea, Jr., was elected executive vice president.

Six-cylinder White Super Power engine. (Source: White Motor Co.)

Redesigned twelve-cylinder "pancake" engine. (Source: White Motor Co.)

World War II

In May 1941, 3,000 White scout cars were delivered to the U.S. Army. White developed the Half-Trac personnel carrier, which performed splendidly and became the standard for the U.S. Army. Great numbers of these Half-Tracs were used along with Army tanks in the pursuit of General Rommel across the Sahara Desert in World War II. Production for 1941 was 17,866 vehicles, nearly all military vehicles.

During 1942-1945, White was committed to building only military vehicles. Scout cars, Half-Tracs, Prime Movers, Tank Destroyers, and cargo trucks were delivered to the U.S. Army well on schedule. In 1942 White was awarded the Army-Navy "E" for excellence in production. Production that year was 18,994 vehicles. During the ensuing war years, White received the second and third Army-Navy "E" stars for continued

White 6-ton 6x6. (Source: White Motor Co.)

excellence in production. In 1943 White was building the M-16 mobile multiple-anti-aircraft gun carriers. In 1944 White began production of the "444" (4-ton, four-wheel-drive truck-tractors) for the U.S. Army, along with the other military vehicles.

Because White had so consistently met its war production schedule, the War Production Board permitted White Motor Company to build 6,398 commercial vehicles for civilian use during the first six months of 1945. During the post-WWII years, White became the foremost leader in the manufacture of motor trucks.

The Evolution to Modern Day

In the early 1950s, by an arrangement with Packard Motor Car Company, White used the 245-cu.-in. Packard six-cylinder engine in its light delivery trucks.

White four-wheeled scout car designed and built in collaboration with the Ordnance Department during WWII. (Source: White Motor Co.)

White military trucks in Africa in 1944. (Source: White Motor Co.)

In 1951 White purchased the assets of the Sterling Motor Truck Corporation of Milwaukee, Wisconsin. In 1953 they sold the Milwaukee Sterling plant and purchased the assets, name, and good will of the Autocar Company, and continued to build Autocar trucks. In 1957 White Motor Company purchased the Reo Motors Incorporated of Lansing, Michigan.

In 1958 White pioneered the use of fiberglass in the production of truck cabs. They also purchased the Diamond-T-Motor Car Company of Chicago, Illinois. The name was changed to Diamond-T-Motor Truck Company, a subsidiary of White Motor Company. Diamond-T and Reo continued to operate as separate entities until 1967 when they were merged to build the Diamond-Reo at the Lansing, Michigan, plant.

Windsor T. White died in 1958 at age 91, at the home of his brother Rollin in Hobe Sound, Florida. Rollin H. White died in 1962.

In 1960 the White Motor Company bought the assets, name, and good will of the Oliver Corporation, which became the White Farm Equipment Company of the White Motor Corporation. During this time the board of directors of White Motor Company sought out Semon E. Knudsen and elected him chairman of the board and chief operating officer on May 1, 1971. Under Knudsen's leadership, White Motor Company

Semon E. Knudsen. (Source: Semon E. Knudsen Estate)

introduced a completely new line of modern, high-horsepower farm tractors and a "New Family of Trucks." The company's truck lines included the White, Autocar, Diamond-Reo, and Western Star nameplates. Remarkable economies were accomplished by engineering innovations that provided flexibility and commonality of components across entire truck lines. Again, success followed "Bunkie" Knudsen, as the White Motor Company prospered, building some of the finest White trucks and farm equipment made in America. Knudsen retired from White Motor Company on July 31, 1979, leaving the company in a very sound financial condition.

After Knudsen's retirement the White Motor Company truck operations were acquired by the Volvo Heavy Truck Corporation, and operated as the White Truck Division of the Volvo-White Truck Corporation until 1989, when White became a division of the Volvo-GM Heavy Truck Corporation. The farm equipment and White-Westinghouse appliances are currently manufactured by White Consolidated Industries.

Source: Horseless Age

Albert Pope

Albert Pope should be credited for organizing perhaps the first and one of the greatest industrial combines. It should have also served as an ominous warning of the folly in over-expansion and over-extension into many geographical areas.

* * *

Albert Augustus Pope was born in Brookline, Massachusetts (a suburb of Boston), on September 20, 1843. Early in life he displayed the dominant personality and ready faculty for organization that later enabled him to become a veritable giant in the industrial world. His early ventures into the business world were during school vacations. His later experiences while a clerk at the Brooks and McCuen Company of Boston, dealing in shoe and leather findings, gave Albert Pope a taste for organization and accomplishment which expanded into a great motivating force, under the influence of his great and compelling enthusiasm.

The outbreak of the Civil War interrupted his civilian endeavors, but Pope turned this adversity into an advantage. Pope enlisted in the Massachusetts Volunteer Militia on August 22, 1862. He was soon commissioned a Second Lieutenant, and a year later was promoted to First Lieutenant. On April 1, 1864, he earned the rank of Captain of his company, and thereafter so distinguished himself in valorous service, that he was brevetted a Major and later promoted to Lieutenant Colonel for gallant conduct.

Pope Manufacturing Company and the Bicycle Industry

After the end of the Civil War, Pope returned to his former employer for a short while, but soon went into business for himself. His first venture was in the manufacture of ornaments for ladies' slippers and as a supplier to shoe manufacturers. On September 20, 1871, he married Miss Abbey Linder of Newton, Massachusetts, and they made their home in that suburb of Boston. In 1876 Pope organized the Pope Manufacturing Company, and proceeded in a modest way the manufacture of various patented novelties.

It was at this time that the seeds of the bicycle industry boom had sprouted and taken root, and would soon grow to astounding proportions. Pope became interested in rider-propelled vehicles, and in the fall of 1877 contracted W.S. Atwell, a local machinist, to build such a vehicle. The cost was $313, and it was a marvelous mechanical achievement in many ways. It definitely convinced Pope of the possibility of enlisting the enthusiasm of other people in the bicycle, and a profitable enterprise emerged. For a short time, Pope imported bicycles from England and merchandised them through the Pope Manufacturing Company. But in 1878, Pope placed an order with the Weed Sewing Machine Company of Hartford, Connecticut, for the production of bicycles in quantities. Ultimately, the demand for bicycles grew to such an extent that Pope Manufacturing Company purchased Weed's manufacturing facilities. At the time Pope had been manufacturing the Columbia Bicycle in his Chicago plant.

At first Pope had to battle for industrial existence. However, through his own efforts he conquered opposition and created a bicycle market by obtaining the use of public parks in Boston, New York, and Chicago for bicyclists, and established for them a measure of genuine recognition. Pope might have left the greater work for others, but he was fired with characteristic zeal. He found common roads hazardous to cyclists because of their poor quality, so he undertook the prodigious task of starting a general movement for their improvement. He enlisted the active support of 45 state governors, and earned the credit for a great number of favorable legislative acts that paved the way for the "Good Roads Movement" for automobiles. That was the field performance of a great leader.

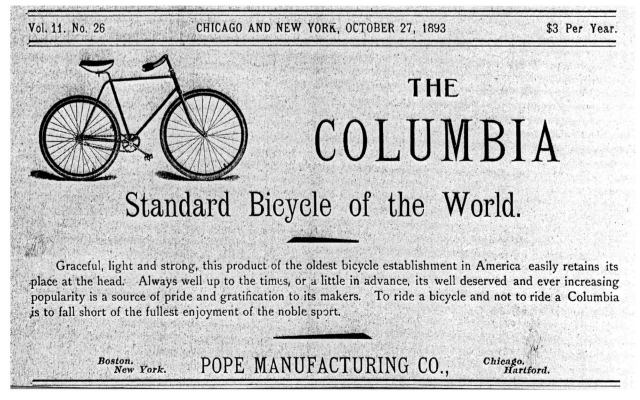

Ad from The Referee *cycle trade journal, 1893.*

The management of the Pope Manufacturing Company was conducted on a "no lesser" basis, as the subsequent growth of the company proved. Pope's absolute domination of the bicycle industry arose largely from his foresight in taking an early and firm grip of the patents situation. He obtained practically every patent on bicycles and their component parts, including the rights to the crank-pedal patent of Pierre Lallement. Through the service of an exceedingly efficient legal department, he increased his company's holdings. For a number of years Pope Manufacturing Company collected a royalty of $10 for every bicycle manufactured and sold in the United States. Not only were these royalties collected scrupulously, but an effective surveillance was maintained over the other bicycle makers, and upon the slightest indication of possible infringement, positive and effective legal action was taken.

It was interesting to note that the first bicycle company to break away from the legal bondage and monopoly of the Pope Manufacturing Company was the Gormully and Jeffery Company, producers of the famous Rambler American bicycle of the early 1880s. Initially, Gormully and Jeffery arranged for a license by the Pope Manufacturing Company until 1886 when R. Philip Gormully issued a declaration that the Gormully and Jeffery Manufacturing Company would no longer pay the royalties. Pope answered with eight lawsuits which eventually were struck down by the courts. Gormully and Jeffery possessed no less than 18 patents on bicycle components registered in their names. Because litigation discouraged Thomas B. Jeffery from pursuing the manufacture of bicycles, and also due to the untimely death of Gormully, Jeffery sold the Gormully and Jeffery bicycle interests to the American Bicycle Company. At that time Gormully and Jeffery Manufacturing Company was the second largest bicycle manufacturer in the United States.

American Bicycle Company

In 1899 the American Bicycle Company was organized as the result of a merger of the A.A. Pope and the Albert G. Spaulding (sporting goods manufacturer) interests. In desperation many bicycle builders sold their assets and manufacturing facilities to the American Bicycle Company, which grew into a consolidation of 45 bicycle manufacturing firms, including the Sterling Bicycle Company of Kenosha, Wisconsin, Indiana Bicycle Company of Indianapolis, the Toledo Bicycle Company, and many others.

By September 1900 the American Bicycle Company stopped production of bicycles at the Toledo plant, and devoted its entire production facilities to the manufacture of Toledo steam vehicles. While the Waverly Electric was originally built by the Indiana Bicycle Company, it became the Waverly Automobile Department of the American Bicycle Company after the consolidation in 1900. The Toledo steam vehicle plant was also administered by the Automobile Department of American Bicycle Company.

Due to the lack of sales, many bicycle plants were idled and A.A. Pope suffered a great financial loss at the turn of the century. The Sterling Bicycle Plant in Kenosha, Wisconsin, was sold to Thomas B. Jeffery in 1900. On August 30, 1902, the American Bicycle Company went into the hands of a receiver, due to the collapse of the bicycle fad and overcapitalization.

Electric Vehicles

In the meantime many things were happening in Philadelphia, New York, New Jersey, and Connecticut. In January 1897, Henry G. Morris and Pedro G. Salom organized the Electric Carriage and Wagon Company to operate and service electric-powered cabs in New York. Morris and Salom had previously designed and built the "Electrobat," which won a gold medal on Thanksgiving Day 1895 in the Chicago Times-Herald race for design reliability. In late 1897, Morris and Salom sold their company to Isaac L. Rice, president of the

1895 Morris and Salom "Electrobat." (Source: Horseless Age)

Electric Storage Battery Company (Exide) of Philadelphia, and the owner of the Charles F. Brush patent on lead-acid storage batteries, and it became the Electric Vehicle Company.

In 1899 William C. Whitney acquired the Electric Vehicle Company from Isaac L. Rice, with the intent of setting up electric cab franchises in all the major cities in the United States. About the same time the Electric

1897 Waverly electric. (Source: Smithsonian Institution)

1897 Waverly limousine.
(Source: Smithsonian
Institution)

Vehicle Company had also acquired the Selden patent rights for the gasoline-powered automobiles. Through a series of mergers the Electric Vehicle Company gained control of the electric vehicle industry. In late 1899 the Riker Electric Vehicle Company also became a unit of the Electric Vehicle Company. A.L. Riker soon left the company after a brief, but stormy, association. George H. Day, former vice president, was elected president of Electric Vehicle Company.

The demand for the electric-powered cabs in all the major cities became so great that the Pope Manufacturing Company of Hartford, Connecticut, was asked to manufacture these vehicles in great quantities. Albert Pope employed some of the best scientific minds in the United States in his research activities.

In 1897 Pope hired Hiram Percy Maxim (son of Hiram S. Maxim, inventor of the machine gun, self-propelling torpedo, and other war implements) as head of the motor carriage department. Maxim was given the responsibility of developing a gasoline-engine-powered vehicle. The result was the Columbia, introduced at the Philadelphia Auto Show in January 1901.

While the Pope Manufacturing Company built about 2,000 Columbia electric cabs, they were very expensive and not profitable for the Electric Vehicle Company. They decided it would be more lucrative to enforce the terms of the Selden patent, citing the success of Pope's bicycle trust royalties.

Hiram Percy Maxim. (Source: Smithsonian
Institution)

1901 Columbia electric. (Source: Smithsonian Institution)

1902 Columbia electric truck. (Source: Smithsonian Institution)

Pope Focuses on Automobiles

The Pope Manufacturing Company was officially incorporated under the laws of the state of New Jersey on February 27, 1903, to take over the consolidation of the American Bicycle Company and the American Cycle Company. Immediately thereafter Pope turned his attention and efforts to the manufacture and sale of

1901 Electric Vehicle Company's engineers, left to right: John W. Chapman, James J. Jones, Herbert Vanderbeck, Ed Tidlund, Hiram P. Maxim (chief engineer), Fred S. Chapman, D.E. Sweetson, Henry Baldwin, Herbert Walden, and Fred A. Law. (Source: Smithsonian Institution)

automobiles. The Toledo steam and the Waverly electric became departments of the Pope Motor Car Company (a new designation for the company subsidiary). New designs were developed for other Pope cars to be made in Hartford, Connecticut, Hagerstown, Maryland, and Hyde Park, Massachusetts.

Since Albert Pope never did things in a small way, with the Columbia Automobile Company and the Waverly and Toledo departments already into automobile production, Pope's restlessness prompted him to acquire the controlling interest in the Robinson Motor Vehicle Company of Boston, Massachusetts, in 1902.

The Robinson Vehicle was the successor to the Bramwell-Robinson Sociable, a three-wheeled vehicle

Bramwell-Robinson Sociable.
(Source: Horseless Age)

1903-04 Pope-Robinson.
(Source: Horseless Age)

powered by a gasoline engine, introduced in May 1899. Bramwell-Robinson Company of Hyde Park, Massachusetts, was founded in 1899, with W.C. Bramwell, inventor, and John T. Robinson, manufacturer of paper box machinery, as partners. After Bramwell left the company, Robinson continued building the Robinson vehicle (a four-wheeled, gasoline-engine-powered carriage) until 1902.

The Pope-Robinson Company was incorporated in September 1902, and the Pope-Robinson car, which succeeded the Robinson vehicle, was introduced in January 1903, and built in Hyde Park, Massachusetts, with a price tag of $6,000. The Albert Pope and John Robinson association was not entirely happy, with Robinson as president, and Edward W. Pope (Albert's nephew) assigned as secretary-treasurer of the Pope-Robinson Company.

The Pope-Robinson car was built only in 1903 and 1904, both years using the same 24-hp, four-cylinder engine, and available in a five-passenger touring car model. For 1904 the wheelbase was increased to 95 in. and the price was reduced to $4,500. Even so the sales of the Pope-Robinson were disappointing. With the sudden death of John Robinson in November 1904, the production of the Pope-Robinson car was discontinued, and the Hyde Park, Massachusetts, plant was sold.

While the Riker (electric and gasoline) vehicles were built in Elizabeth, New Jersey, and the Columbia (electric and gasoline) cars were built in the Pope Manufacturing Company plant in Hartford, Connecticut, they were merchandised through the Electric Vehicle Company of Hartford, Connecticut, the Pennsylvania Electric Vehicle Company of Philadelphia, Pennsylvania, the Washington Electric Vehicle Transportation Company of Washington, D.C., the Columbia Motor Vehicle Company of Boston, Massachusetts, and A. Bianchi in Paris, France. Their gasoline engines of 4, 8, and 16 hp could propel the vehicles from 20 to 40 mph. Their electric vehicles could travel up to 40 miles on each charge of the Exide batteries. The Riker vehicle was discontinued in 1903, but Pope continued building the Columbia vehicles.

Pope then began, in rapid succession, the development and manufacture of Pope-Hartford and Pope-Tribune gasoline-engine-powered cars. A new vehicle was designed and tested in 1903: the Pope-Hartford car powered by a 10-hp, single-cylinder, gasoline engine. Introduced in early 1904, it was available in runabout and tonneau body styles, priced at $1,050 and $1,200, respectively.

1904 Pope-Tribune.
(Source: Borg-Warner)

1905 Pope-Tribune Model 4 at Hagerstown, Maryland, factory. (Source: National Museum of American History, Smithsonian Institution)

For 1905, both 10-hp, single-cylinder and 16-hp, two-cylinder engines were offered in the tonneau models only, priced at $1,000 and $1,600, respectively. The tonneau with the two-cylinder engine had a wheelbase of 88 in., 10 in. longer than the one powered by the single-cylinder engine.

The Pope-Tribune was, in fact, a downsized Pope-Hartford, introduced in 1904 as an economical runabout priced at $650. It was built in Hagerstown, Maryland, in a former bicycle factory. Unfortunately, the size of the Pope-Tribune grew over the next three years, and the price escalated to over $2,700. As a result, sales dropped and production was terminated in 1908.

In 1906 and 1907 Pope-Hartford cars were available with both the two-cylinder engine, and the new four-cylinder gasoline engine introduced in January 1906. The two-cylinder-engine-powered models were discontinued after 1907. The Pope-Hartford car prices escalated, and the proliferation of body types available did not make for a profitable operation, with a production of only a few hundred cars per year.

In February 1907 the Electric Vehicle Company introduced the Columbia Mark LXVI electrically driven

1905 Pope-Hartford Model B (at right). (Source: National Museum of American History, Smithsonian Institution)

gasoline-engine-powered car, using the Justice B. Entz design power transmission made by the General Electric Company. The car was powered by a four-cylinder vertical gasoline engine of 5-in. bore and 5-in. stroke, developing 45 hp. The engine crankshaft was connected to an external field frame of the General Electric direct current generator. The generator electrical output in turn energized an electric motor, whose armature was connected by a propeller shaft and universal joints to the rear axle driving gears. Controls for the output of the unit and the speed of the vehicle were on the steering column.

Pope is Reorganized

The Pope Manufacturing Company, after vain attempts to raise cash to tide over its financial crisis, went into the hands of a receiver at Hartford, Connecticut, on August 14, 1907. Egbert J. Tambly, a Newark, New Jersey, attorney, and Albert L. Pope (A.A. Pope's eldest son) were appointed receivers. On the same day application was made for the appointment of Albert L. Pope as receiver of the Pope-Tribune's Hagerstown, Maryland, plant. Also on August 14, 1907, the Pope Motor Car Company of Toledo, Ohio, went into bankruptcy, with liabilities of over $500,000 and paper assets of about $2,000,000. Albert L. Pope, president of the company, was named receiver of that as well.

Pope Manufacturing Company ignored the warnings of 1903 and the investment market rated the enterprise accordingly. The vicious "cumulative dividend" proviso, which made one year's unpaid dividend a contingent charge on the company's future revenues, had compelled the eventual receivership and reorganization of the Pope Manufacturing Company, in view of the financial panic of 1907.

A new Pope Manufacturing Company was incorporated on December 12, 1908, under the laws of the state of Connecticut. The sale of the Pope-Waverly plant at Indianapolis was completed on September 25, 1908, when Judge A.B. Anderson of the United States District Court directed Albert L. Pope to accept the offer of $200,000 made by Herbert H. Rice and Wilbur C. Johnson for the property, merchandise, patents, good will,

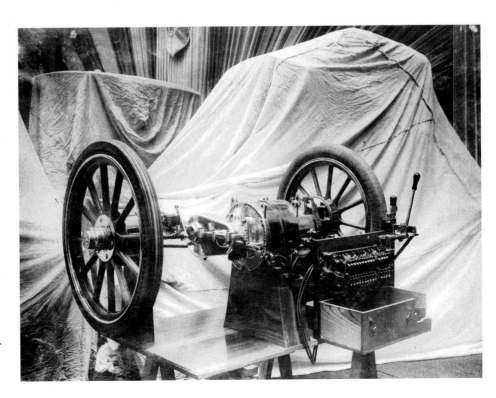

1910 Waverly electric drive.
(Source: Smithsonian Institution)

and business. The new owners organized as the Waverly Company under the laws of the state of Indiana, with a capitalization of $225,000. The company was operated by William B. Cooley, president, Herbert H. Rice, vice president, Carl VonHake, treasurer, and Wilbur C. Johnson, secretary, continuing in the positions they occupied under the Pope regime. They continued building electric-powered passenger vehicles and trucks through 1908.

On April 1, 1909, John N. Willys heard that the Pope-Toledo plant was up for sale. Willys arrived in Toledo on Friday morning April 2nd, and with his characteristic drive, he visited the plant in the forenoon and met with Albert L. Pope in New York on Saturday, April 3rd. He obtained an option on the property on Monday, April 5th by a deposit of $25,000. The reported selling price was between $350,000 and $400,000. Only 48 hours elapsed between the start of the negotiations and the obtaining of the option. Not only did the Toledo plant provide Willys with approximately 3,000,000 sq. ft. of manufacturing area and production facilities he needed, but it was also one of the newest and best-equipped automobile plants in the United States.

On June 14, 1909, Judge Cross of the United States Circuit Court in Trenton, New Jersey, issued an approval and instructions for the payment by the reorganization committee for the assets and properties of the Electric Vehicle Company, including the Selden patent rights. This transaction resulted in the reorganization of the Electric Vehicle Company, the new corporation being called the Columbia Motor Car Company. It had been reported that the committee had paid $300,000 for the Electric Vehicle Company's assets and properties.

The Columbia Motor Car Company was incorporated under the laws of the state of Connecticut on June 30, 1909, and capitalized at $3,000,000. Herbert Lloyd (a former large stockholder of Electric Vehicle, and president of the Exide Storage Battery Company) was elected president. Henry W. Nuckols (a receiver of Electric Vehicle Company since December 10, 1907) of Hartford, Connecticut, was elected vice president, treasurer, and general manager. The branch offices of the Electric Vehicle Company were discontinued and supplanted by distributing agencies.

A.A. Pope Dies

The crushing blow to the crumbling Pope empire came on Tuesday, August 10, 1909, by the sudden and untimely death of Colonel Albert Augustus Pope at his estate in Cohasset, Massachusetts, at the age of 66. While much of Pope's personal wealth was carried away with the bankruptcy of the American Bicycle Company, he was said to have retained a considerable estate, reported to have been in excess of $2,000,000. The responsibility of administrating the Pope enterprises fell on the shoulders of his eldest son, Albert, president of the Pope Manufacturing Company. Three other children survived A.A. Pope: Harold L. Pope, Ralph L. Pope, and daughter Mrs. Freeman L. Hinckley.

Picking Up the Pieces

A year of retrenchment followed for the Pope Manufacturing Company in 1910. The Pope-Robinson plant in Hyde Park, Massachusetts, had previously been sold in 1904, the Pope-Waverly plant in Indianapolis was sold to H.H. Rice and Wilbur Johnson on September 25, 1908, the Pope-Tribune plant in Hagerstown, Maryland, was sold in November 1908, the Pope-Toledo plant in April 1909, and the Electric Vehicle Company (Columbia) was disposed of on June 14, 1909. Thus, the only vehicle being manufactured by the Pope Manufacturing Company was the Pope-Hartford. The bicycle business was in total disarray as the bicycle fad collapsed, and the plants were idle.

1910 Pope-Hartford T roadster. (Source: Pope Catalogue)

Banking on the Pope-Hartford

In an attempt to stir buying interest, the 1911 Pope-Hartford cars became bigger and more powerful. The Pope-Hartford Model T was available in six body types on a 122-in. wheelbase chassis, powered by a 40-hp, four-cylinder engine. However, the price had escalated to $2,750 for the roadster and as much as $4,750 for the landaulet.

Pope also indulged in automobile racing to prove the performance of Pope-Hartford cars. The four-day speed carnival on the Atlantic-Pablo Beach at Jacksonville, Florida, came to a climax on Friday, March 31, 1911, when Louis A. Disbrow won the 300-mile race driving a Pope-Hartford racing car; he covered the distance in 3 hours, 53 minutes, and 33.50 seconds, at an average speed of over 77 mph. He also established new speed records at 150, 200, and 250 miles. Disbrow had previously set a record of 81.65 mph for 1 mile on March 28, 1911, and for 10 miles in 7 minutes, 42.39 seconds on Thursday, March 30, 1911.

Two Pope-Hartford racing cars were entered in the Indianapolis 500. Pope-Hartford #5, driven by Louis Disbrow, collided with Teddy Tetzlaff's Lozier on the main straightaway on lap 45, and Frank Fox driving the Pope-Hartford #6 was awarded 22nd place after being flagged at the conclusion of the race. These were the only attempts by Pope-Hartford at the Indianapolis Motor Speedway.

While it had been known for some time that Pope Manufacturing-Company had a six-cylinder-engine-powered model in the works, the new Pope-Hartford Six Model Y and the Pope-Hartford Four Model W were introduced in late 1910. The cars were similar in design, powered by 44.8- and 36.1-hp (ALAM) engines, respectively. The frame construction was also similar, except the Y had a 10-in. longer wheelbase to accommodate the six-cylinder engine.

Three new designs followed the trend of the times in touring car design. The five-passenger model termed the "torpedo" version, while lacking the outlandish construction of competitive torpedo designs, had relatively high sides and an overhanging dash. The seven-passenger model was of the closed-front variety. The four-passenger small tonneau model possessed the seat lines of the touring, but was more compact, and in place of the front doors had a deep scuttle dash whose ends extended into the body sides. Six body types were available in the W and four body styles in the Y. The price range was from $3,000 for the W touring to $4,150 for the W limousine, and $4,000 for the Y touring to $5,150 for the Y limousine and landaulet.

Disbrow driving #5 Pope-Hartford in the 1911 Indianapolis 500. (Source: Indianapolis Motor Speedway Corp.)

For 1912 the Pope-Hartford models continued unchanged, except the model designations were now Model 27 for the four-cylinder cars and Model 28 for the six-cylinder cars, with an engine output of 44.6 hp (ALAM). A new closed-body type called the Berline was added to both car lines, priced at $4,400 and $5,400, respectively. The Pope-Hartford proliferated to 17 body types, at a price range from $3,000 to $5,400.

Strangely, the annual report of the Pope Manufacturing Company for the year ending July 31, 1912, showed

Frank Fox in a 1911 Pope-Hartford. (Source: Indianapolis Motor Speedway Corp.)

net earnings of $251,293 on a gross business of $3,734,112. Dividend payments for 1912 amounted to $174,892. A dividend of 2.5% was paid on common stock on July 31, 1910, and a 1% dividend was paid on January 31, 1912. On December 31, 1912, the Pope Manufacturing Company was reincorporated in Massachusetts, with the same capitalization as before, having the outstanding stock exchanged dollar for dollar for stock in the new company. The authorized capitalization was $4,000,000 common stock and $2,500,000 6% preferred. The officers of the company at that time were: Albert L. Pope president, C.E. Walker first vice president, W.C. Walker secretary, and George Pope (A.A.'s cousin) treasurer. In addition to the company officers, Edward W. Pope (son of George Pope) was on the board of directors.

George Pope, generally known as "Colonel George," was one of the organizers of the automobile show movement. He was chairman of the show committee of the Association of Licensed Automobile Manufacturers (ALAM), remaining through their successive transformations. In that capacity he had active charge of the New York auto shows from 1906 to 1913, subsequently retaining chairmanship of the show committee of the National Automobile Chamber of Commerce (NACC). He had been treasurer of the NACC since its organization.

Colonel George Pope. (Source: Horseless Age)

The 1913 Pope-Hartford consisted of three lines of cars, identified as Models 31, 33, and 29. Models 31 and 33 were powered by a four-cylinder engine, the 31 having a shorter wheelbase of 118 in. Model 29 superseded Model 28 unchanged, except for a price increase to a range of $4,250 to $5,500. The Models 31 and 33 had a price range of $2,250 to $4,550. The proliferation of body types to 18 and limited car sales caused a deficit of $181,780 for 1913. On October 27, 1913, the Pope Manufacturing Company was placed in receivership by Judge Joseph P. Tuttle, appointing George Pope as receiver, under bond of $200,000. George Pope issued a statement through his counsel, giving the cause of the receivership as due to the restricted credit, owing largely to the maturity of $1,000,000 notes in April.

While the company's plans were already in place, and a complete stock of parts were ordered, it is not known how many cars were actually built in 1914. The Pope-Hartford line was reduced to one model (the 35), a continuation of the previous year's Model 31. In the meantime the properties and assets were being sold off piecemeal. In the final sale on January 2, 1915, the Pope Manufacturing Company main plant in Hartford, Connecticut, was sold to the Pratt and Whitney Company.

Lessons Learned

There were many lessons to be learned from the saga of the Pope Manufacturing Company. Had the dangers of over-expansion been studied, it might have spared the future demise of the United States Motor Company, and the grief of the Willys-Overland Company and the Studebaker Corporation.

The Columbia Motor Car Company was doing fairly well, being one of the first American cars to offer the Knight sleeve-valve engine, until it "jumped out of the frying pan into the fire" by joining the United States Motor Company in early 1910. General Motors Company survived and was spared this agony by the banking institutions coming to their rescue in 1910 and 1920. The independents of the 1950s should have reviewed and studied Pope's history, before indulging in their hasty mergers.

While the Pope Manufacturing Company enterprises failed, their former plants and facilities continued building constructively, contributing to the economy of the United States of America.

Henry Leland

Few people in the automobile business were more genuinely admired and respected than Henry Leland. He was the bulwark of honesty and good faith, and was a powerful influence for precise manufacturing and honest dealing for over twenty years. In retrospect it is ironic that the government of a country for whom Henry Leland had done so much should deal him such a cruel and unfair blow that helped put him out of business.

* * *

Henry Martyn Leland was born in Danville, Vermont, on February 16, 1843. Henry's family moved to Millbury, Massachusetts, shortly after he was born. He was employed part-time while still in school, and when he was only nine years old he invented the "hair clipper" for his employer in 1852. While his employer profited greatly from clipper sales, Leland received no compensation other than a 50 cent increase in his pay envelope.

Leland started his mechanical trade training when only 13, at the Knowles Crompton Loom Works of Worcester, Massachusetts, and completed his apprenticeship at age 19. By 1861 he was turning gun stocks at the United States Arsenal in Springfield, Massachusetts. He later became a toolmaker at the Samuel Colt Arms factory in Hartford, Connecticut.

On September 25, 1867, Henry Leland married Ellen Hull, daughter of Elias Hull of Worcester, Massachusetts. Their daughter Gertrude was born on June 8, 1868; Wilfred Chester Leland, their son, was born on November 7, 1869; and Edith Miriam Leland came many years later on February 2, 1882.

Robert C. Faulconer (left) and Henry Leland in their office at Leland and Faulconer Manufacturing Company. (Source: Leland Collection)

On July 1, 1872, Leland joined the Brown and Sharpe Tool Works in Providence, Rhode Island. After working inside the plant for several years, he was named special representative of the famous tool and instrument manufacturer, traveling to most of the industrial cities of the United States.

Leland's First Venture

After several years of traveling as a machine tool representative, he decided to relocate to the Midwest. He had chosen Chicago, but upon arriving there on May 1, 1886, as he was about to leave the station, he was confronted by a riot at Haymarket Square. The police ranks trying to maintain peace at a meeting called by labor leaders were attacked with a bomb which killed seven policemen and wounded numerous others. Leland turned around, walked to the nearest ticket window, and asked for a ticket on the first train out. Thus, that evening he found himself bound for Detroit. His plans were vague; however, when he stopped in Detroit to rest from the train ride he was pleased with what he saw and decided to remain there. Within a few days he obtained quarters in a building loft, and went to work in earnest. It wasn't long before he teamed up with Robert C. Faulconer, a successful lumberyard owner, to establish a company for the manufacture of precision machinery and tools, located at 96 Bates Street in Detroit, known as the Leland and Faulconer Manufacturing Company.

Wilfred C. Leland came to Detroit in 1890 on his 21st birthday, and joined his father and Robert Faulconer, who were already established as tool precisionists. Ernest E. Sweet, a skilled machinist, also joined the

Leland and Faulconer Manufacturing Company employees. Leland is third from right in second row. (Source: Leland Collection)

Leland and Faulconer Manufacturing Company. On September 19, 1890, Charles H. Norton, a brilliant tool design engineer whom Leland had known from the days of the Brown and Sharpe Tool Works, joined the Leland and Faulconer Company as a full-fledged partner. Norton stayed with the company through the growing years, until 1896 when he decided to move to Bridgeport, Connecticut, where he established the Norton Company, manufacturer of the famous "Norton Grinder."

How Cadillac Began

While the first plans to organize began in late 1899, Wayne County incorporation records indicate that the Detroit Automobile Company was incorporated as of January 1, 1900, with the actual filing date on February 14, 1900. The Articles of Incorporation were signed by Frank R. Alderman, secretary, notarized by Charles D. Willis. The purpose given as the reason for incorporation was "Preparing for the manufacture of automobiles."

The officers of the company were listed as follows: Clarence A. Black president, Albert E. White vice president, Frank R. Alderman secretary, William H. Murphy treasurer, Henry Ford mechanical superintendent.

The board of directors consisted of: C.A. Black, A.E. White, F.R. Alderman, W.H. Murphy, F.W. Eddy, Lem W. Bowen, and Henry Ford. The addresses given were: Union Trust Building, Detroit, Michigan, Cass Avenue at Michigan Central Railroad, #40 Moffatt Building, Detroit, Michigan.

The stockholders and their holdings were listed as follows: Clarence A. Black – 100 shares, Albert E. White – 100 shares, William H. Murphy – 100 shares, Frank R. Alderman – 100 shares, Frank J. Haecker – 100 shares, William C. McMillan – 100 shares, Patrick A. Dacey – 100 shares, Henry Ford – 100 shares, Mark Hopkins - 100 shares, Edward A. Leonard - 50 shares, Benjamin R. Hoyt - 50 shares, Lem W. Bowen - 50 shares, Stafford D. DeLano -50 shares, Dora H. Black - 50 shares, William H. Lake - 50 shares, William C. Maybury - 50 shares, Ellery I. Garfield - 50 shares. The capital stock was given as $150,000 with $15,000 paid in.

Up to that time Henry Ford had built only one small car, which had the Pennington engine design and small wire spoke wheels. After the Detroit Automobile Company was organized, two cars were under construction: a two-seat, four-passenger car with wood spoke wheels, and a single two-passenger runabout with wire spoke wheels. Neither of these cars was completed when the company found itself in financial trouble in 1901.

The company was reorganized on November 30, 1901, and became the Henry Ford Company until March 10, 1902, at which time Henry Ford resigned. Part of the resignation terms were that Ford retain the racing car and the designs.

During the subsequent reorganization on August 22, 1902, the name was changed to Cadillac Automobile Company, with Lem W. Bowen (one of the original Detroit Automobile Company backers) named president, and Clarence A. Black vice president. Oliver E. Barthel succeeded Henry Ford as superintendent of the mechanical department. The first Cadillac car was completed on October 17, 1902, and made its first public appearance at the National Automobile Show in New York in January 1903. William E. Metzger, who was named general sales manager in late 1902, obtained orders for 2,700 Cadillac cars at the show, even before it was formally introduced, and actually only three cars had been built up to that time. By the end of 1903, 3,000 Cadillac cars were built and sold. Metzger remained with Cadillac until 1908, when he resigned to join the E.M.F. organization.

Henry Leland and his grandson Wilfred Leland, Jr.
(Source: Leland Collection)

Leland Hooks Up With Cadillac

The Leland and Faulconer Company at 96 Bates Street was located almost directly across the block from the Charles A. Strelinger and Company at 86 Woodward Avenue, the foremost supplier of hardware, tools, and related manufacturing needs. Leland's and Strelinger's association and friendship grew through many years. It also happened at the time that an aggressive young engineer named

Edith Miriam, Henry, Wilfred, and Wilfred Jr., left to right. (Source: Leland Collection)

Alanson P. Brush was employed by the Charles A. Strelinger and Company. Brush was working and developing a successful single-cylinder gasoline-fueled engine. At this time the Leland and Faulconer Company was manufacturing the transmissions for Olds Motor Works.

When Cadillac reorganized in 1902, Leland and Faulconer Company was called upon to furnish the engine and transmission for the new Cadillac car. They manufactured the transmission and the single-cylinder engine of 6.5 hp, designed by Alanson Brush. Brush joined the Leland and Faulconer Company in 1902 and stayed with them through 1904.

Oliver E. Barthel stayed on as chief engineer and superintendent of the mechanical department of Cadillac Automobile Company to design a car for production. However, Barthel was also moonlighting in the evenings, working with Henry Ford and C. Harold Wills in the building and completion of Henry Ford's second racing car, the famous "999." Eventually, when the Cadillac management found out about this, Barthel was relieved of his duties at Cadillac, and was succeeded by Pat Hussey. Barthel stayed with Henry Ford through the organization of the Ford Motor Company, and until 1904 when he left to establish a consulting engineering firm, doing design and engineering work for many automobile manufacturing companies.

On April 13, 1904, a serious fire broke out in the Cadillac plant, causing $240,000 in damage. Martin Gorman, foreman of the frame department, sustained serious injuries in attempting to extinguish the blaze, and died a few days later. The Cadillac Model B was shipped in lieu of Model A starting on May 7, 1904.

During 1904 Cadillac Automobile Company had its growing pains, including many organizational squabbles. Finally, late in 1904, Henry M. Leland was asked to become general manager of Cadillac. Leland, realizing how important it was to save Cadillac in order to have a customer for his engines, agreed on the basis of a merger between the Cadillac Automobile Company and the Leland and Faulconer Company. On a blustery cold December 23, 1904, Henry Leland, his son Wilfred, Ernest Sweet, and Alanson Brush trudged through the snow drifts on Cass Avenue to join the Cadillac Automobile Company. On January 11, 1905, Leland was appointed general manager of Cadillac Automobile Company. Brush became the engine designer for Cadillac, and was credited with the design of Cadillac's new four-cylinder engine in 1905. After reorganization and incorporation in the state of Michigan, it became the Cadillac Motor Car Company on October 27, 1905.

Despite the rough going, and with the help of the excellent leadership of Henry M. Leland, Cadillac won wide acclaim for quality and precision workmanship. While Leland surrounded himself with qualified and skilled people, he also was a meticulous and scrupulously precise craftsman due to his experience in arms manufacture. A new four-cylinder engine was introduced in 1905. The year 1905 was profitable for Cadillac, as they sold 4,029 single-cylinder engine models, and 156 cars powered by the new four-cylinder engine. Brush proved his engineering skill in the design of Cadillac's four-cylinder engine, but left Cadillac in late 1906 to form his own automobile company. (See the Brush chapter in Volume 4.)

Cadillac	Wheelbase (in.)	Price	No. of Cylinders	Bore and Stroke (in.)	hp (adv)	hp (ALAM)
1903 Model A	72	$750-850	1	5.00 × 5.00	6.5	10
1904 Model A	72	$750	1	5.00 × 5.00	6.5	10
1904 Model B	76	$900	1	5.00 × 5.00	8.25	10
1905 Model B	76	$900	1	5.00 × 5.00	9	10
1905 Model C	76	$950	1	5.00 × 5.00	9	10
1905 Model E	76	$750	1	5.00 × 5.00	9	10
1905 Model F	76	$950	1	5.00 × 5.00	9	10
1905 Model D	100	$2,800	4	4.37 × 5.00	30	30.6
1906 Model K	74	$750	1	5.00 × 5.00	10	10
1906 Model M	76	$950	1	5.00 × 5.00	10	10
1906 Model H	102	$2,500	4	4.37 × 5.00	30	30.6
1906 Model L	110	$3,750	4	5.00 × 5.00	40	40
1907 Model K	74	$800	1	5.00 × 5.00	10	10
1907 Model M	76	$950	1	5.00 × 5.00	10	10
1907 Model G	100	$2,000	4	4.00 × 4.50	20	25.6
1907 Model H	102	$2,500	4	4.37 × 5.00	30	30.6
1908 Model S	82	$850	1	5.00 × 5.00	10	10
1908 Model G	100	$2,000	4	4.00 × 4.50	20	25.6
1908 Model H	102	$2,500	4	4.37 × 5.00	30	30.6
1909 Model 30	106	$1,400	4	4.00 × 4.50	30	25.6

TO CADILLAC DEALERS

CIRCULAR LETTER NO. 13

AS you know, we have had a fire, a bad one, but not as bad as reported, nor bad enough to put us out of business. Our engine factory, machine and power plant are both in full operation. Our warehouse, 300 x 180, with large quantities of material, including 2000 engines and 200 finished model "A's" is undamaged, This warehouse will be converted into an assembly shop within a week. A large two-story factory near the plant of one of our body makers has a large force of finishers at work on bodies. We have sufficient material coming in to make nearly forty machines per day. The large manufacturers who supplied us with bodies, axles, tires, wheels, frames, chains and other material, have been instructed to duplicate orders already filled. We have lost no tools, dies, jigs, patterns, drawings or special machinery. Our entire force of 600 employees are now at work. Within 30 days we will be shipping model B's. In the meantime we can fill a limited quantity of orders for model A's. If you cannot hold your trade until we can get machines to you, do not hesitate to save your profit by selling another machine, *if you can get it,* but remember that the Cadillac people have a **reputation** for hustling, and we are likely under present conditions to be able to fill your order as soon as any other concern, who make goods of our class. We can make no definite promises at this date. If you cannot wait, please cancel your unfilled orders and we will assign them to those who can We can see no reason why we cannot resume model B shipments in a month. We cannot, of course, fill all our orders in thirty days, but we can begin. Please canvass your trade at once and let us know how many *standard* machines of each model you will need to fill your orders, and we will try to five you some idea of what we can do.

To the many dealers who have expressed their regrets and assurance of support we wish to say, our loss is great but our energy is greatar. We thank you and assure you that we are not discouraged. We feel *your* loss keenly. We know you need the machines and we will have them for you. Give us a little time is all we ask. We are unable to reply promptly to the many expressions of sympathy from our friends. We know you will pardon us. Give us a little time, and we will make Phoenix look like a "plugged nickel."

Yours for a greater business than ever,

CADILLAC AUTOMOBILE COMPANY

P. S. Contract for rebuilding the burned portion of our plant has been let and work begun.

*1905 one-cylinder Cadillac,
their first closed car.
(Source: Leland Collection)*

Henry Leland introduced the concept of interchangeability of parts, made possible by precise manufacturing. Because of this, in 1908 Cadillac became the first American car manufacturer to win the coveted DeWar Trophy, awarded by the Royal Automobile Club of Great Britain for the greatest contribution to the advancement of the automotive industry. Leland was also responsible for the first importation of the famous "Jo-Blocks," the extremely accurate measuring blocks perfected by the Swedish precisionist Carl Edward Johansson.

Cadillac Joins GM

On September 16, 1908, General Motors Company was incorporated in the state of New Jersey, with a capitalization of $2,000,000. The newly formed board of directors increased the capitalization to $12,500,000 on September 28, 1908. The financial figures reflected the magic of William C. Durant's name and confidence in his financial audacity.

On September 29, 1908, General Motors Company purchased the Buick Motor Company, and with Buick as the central core, Durant quickly acquired the Olds Motor Works and the Oakland Motor Car Company.

Cadillac Motor Car Company joined General Motors Company on July 29, 1909. At the time Lem W. Bowen was president of Cadillac, Clarence A. Black vice president, Henry Leland was general manager, William H. Murphy treasurer, and Wilfred Leland secretary. After the merger Henry Leland was elected president of Cadillac, and his son Wilfred was elected vice president and general manager. The General Motors Company was incorporated in the state of Michigan on December 28, 1909.

Wilfred and Henry Leland.
(Source: Leland Collection)

Fisher Closed Body Company

Meanwhile the Fisher Brothers of Norwalk, Ohio, not content with their carriage body business, encouraged brothers Frederick J. and Charles T. Fisher to organize the Fisher Body Company in Detroit on July 22, 1908. They were later joined by brothers William A., Laurence P., Edward F., Alfred J., and Howard A. Fisher. They specialized in and campaigned for closed automobile bodies.

In spite of his normal conservative nature, Henry Leland knew a good thing when he saw it, and in 1910 he rewarded the Fisher Brothers with an order for 150 enclosed automobile bodies (sedans) for Cadillac cars. That was the first time that closed automobile bodies were built on a production basis. Up to that time the only closed car bodies were handcrafted and prohibitively priced limousines, cabriolets, Berlines, and other expensive types produced by the world's finest coach builders. Typical was a Salon Limousine displayed at the St. Louis World's Fair, with a price tag of $18,000.

The Fisher Closed Body Company was organized in 1911. By the end of 1919, 10% of the cars built in the United States were of the closed body type. By the end of 1929, 90% of the cars were built with closed bodies. Thus the famous phrase "Body by Fisher" became a household word during the second and third decades of the 20th century.

The Birth of Lincoln Motor Company

As the result of a disagreement between Henry Leland and William C. Durant in regard to aircraft engines and other military materiel production, Leland and his son Wilfred resigned from Cadillac. Richard H. Collins succeeded Leland as president of Cadillac on July 16, 1917.

Henry, Wilfred, William Murphy, and Joe Boyer made plans for the organization of the Lincoln Motor Company, and agreed to build and equip a small plant to build 14 Liberty aircraft engines per day for the United States Army Air Corps. To finance the venture, they arranged personal loans amounting to $2,000,000. The Lincoln Motor Company was organized and incorporated in the state of Michigan on August 29, 1917.

The Lincoln Motor Company entered into a contract with the U.S. Army Air Corps on August 31, 1917, for the building of Liberty aircraft engines at cost plus 15%. To meet the government's request to increase the production, and to equip the plant to build 70 engines per day, required an $8,500,000 outlay for additional plant space and equipment. To provide the additional capital, a modifying contract dated December 10, 1917, was drawn up and signed by the Lincoln Motor Company and the United States Army Air Corps. In this contract Colonel Montgomery of the U. S. Army Air Corps insisted on a reduced profit of only 12.5%, and for the estimated cost of the engine to be reduced from $6,087 to $5,000 per engine. In this contract the government would allow 40% depreciation on tools and equipment.

First production Liberty aircraft engine built by Lincoln Motor Company. (Source: Leland Collection)

Lincoln purchased the plant site at Warren and Livernois Avenues in Detroit, broke ground on September 21, 1917, and construction work proceeded 24 hours per day, seven days a week. The first machines were installed on October 1st, even before the main buildings were completed, and first machining operations started on November 1st. The main factory buildings were completed on December 25th, and engine assembly started on February 4, 1918. Since Lincoln Motor Company was organized solely to build aircraft engines in the shortest period of time, they did not have to devote time to research and engineering, since this had already been done by Jesse G. Vincent and Elbert J. Hall.

On July 31, 1918, Lincoln Motor Company and the United States Army Air Corps entered into a non-cancellable contract for 9,000 Liberty aircraft engines at a fixed price of $4,000 per engine, with an option that an order for 8,000 additional Liberty aircraft engines may be placed at the same fixed price of $4,000 per engine. The contract also provided that if Lincoln Motor Company were not permitted to build the full amount of 17,000 engines, there should be an additional depreciation and amortization allowance of 40% on the entire plant and equipment, after the depreciation and amortization already provided for had been deducted. By November 11, 1918 (Armistice Day), the U.S. Army Air Corps had accepted 3,798 Liberty engines built by Lincoln Motor Company.

During November 1918, the United States Government advised the Lincoln Motor Company that it wanted to limit the total engine production to 6,500 engines. A supplemental contract was agreed upon and signed by the U.S. Army Air Corps and the Lincoln Motor Company on January 6, 1919. By this contract the

Alvan Macauley explaining the Liberty aircraft engine to Henry Leland in 1917. (Source: P.M.C. Co.)

different rates of depreciation and amortization previously agreed upon were merged into one flat depreciation of 55% on the entire plant and equipment provided for the purpose of this contract.

Production of the balance of the Liberty engines contracted for extended into the middle of 1919. Contract termination arrangements, transfer of the titles of properties to Lincoln, and rearrangements of the plant facilities for automotive production went on during 1919. The comparison Lincoln production of Liberty aircraft engines to other manufacturers was as follows: Lincoln, 6,500 engines; Packard, 6,500 engines; Ford Motor Co., 3,950 engines; General Motors, 2,528 engines; Nordyke and Marmon, 1,000 engines.

Lincoln Enters the Automotive World

In late 1919 Lincoln Motor Company was reincorporated in the state of Delaware for the purpose of automobile manufacturing, with Henry Leland president, Wilfred Leland vice president, and William T. Nash secretary-treasurer.

During the week of January 24, 1920, Leland announced the new Lincoln car to be powered by a new V-8 engine. The design and engineering had been completed, and they were entering the stage of manufacture, planning to build 6,000 cars in 1920. The new Lincoln had many features, refinements, and improvements over its competitors, particularly in the new V-8 engine design; the Lincoln engine cylinders had an included angle of 60 degrees, eliminating most of the inertia vibrations inherent in other V-8 engines.

However, the untimely introduction of the Lincoln in 1920 coincided with the beginning of the period of depression which affected all businesses. The purchase of machinery, equipment, and production material at peak prices placed a great financial burden on the Lincoln Motor Company. A financial crisis was warded off in July 1920 when the Lelands and the board of directors endorsed $5,000,000 in bank notes. A mortgage of $2,500,000 was also authorized, of which the leading directors purchased $1,250,000 of the lien. But in so doing Henry and Wilfred lost control of the B voting stock.

Wilfred C. Leland. (Source: Leland Collection)

Lincoln Crashes

Lincoln car sales amounted to only 285 cars in 1920, and the business climate of 1921 did not improve. The volume of sales anticipated did not develop, as only 804 cars were sold in 1921. This caused a greater drain on the operating capital. Thus, on November 8, 1921, at the board of directors meeting, a petition for voluntary bankruptcy was approved by a vote of 6 to 3. The only directors who voted against the petition were Henry Leland, Wilfred Leland, and William Nash. The assets of Lincoln Motor Company were estimated at $14,300,000 and the liabilities at $8,237,280. The Detroit Trust Company was appointed receiver by United States Judge Arthur J. Tuttle.

Concurrently with the receivership proceedings, on November 28, 1921, the United States Government, through the Collector of Internal Revenue, filed in the office of the Clerk of the United States District Court a claim in the amount of $4,500,000 for additional income and excess profit taxes from the Lincoln Motor Company. Lincoln had already paid $4,126,000 in taxes, the amount arrived at by the United States War Department accounting. However, the United States Treasury Department reopened the case, ignoring the settlement made by the War Department, and proceeded to make a claim for the alleged additional taxes. Filing of the claim by the Internal Revenue Service prevented the Lincoln Motor Company from putting through any refinancing plans, and was one of the principal reasons for the receivership.

On December 29, 1921, the Detroit Trust Company receiver for the Lincoln Motor Company learned from the United States Treasury Department that the claim for the additional taxes would be reduced from $4,505,681 to about $500,000. Unfortunately the news of the United States Government's reduction was received too late in the day to be available at that day's meeting of creditors. Although plans for

1921 Lincoln "Road Runner," Lincoln Highway Pathfinder car. (Source: Leland Collection)

1921 Lincoln cars awaiting shipment. (Source: Leland Collection)

1921 Lincoln Model 104 four-passenger coupe. (Source: Leland Collection, Lincoln Owner's Club)

rehabilitation for the company had been laid out by the receiver, and it was hoped that a reorganization could be made, all attempts failed.

A receivers' sale was held on February 4, 1922, at which Henry Ford bought the Lincoln Motor Company for a reported amount of $8,000,000. The bid by Ford Motor Company was the only one received by that date; two other interested bidders, who had qualified to bid by previously posting certified checks, had failed to submit bids on the day of the sale.

There were certain conditions that the Lelands claim Henry Ford had agreed to before the purchase of the Lincoln Motor Company. In a lengthy letter dated April 10, 1924, from Wilfred C. Leland to Henry Ford, Wilfred charged that the present owner of the Lincoln Motor Company failed to make good his promise to return in full the investment of every qualifying original Lincoln Motor Company stockholder. It was claimed by the Lelands that Mr. Ford paid creditors $4,000,000 only because he was legally obligated to do so. Wilfred Leland did not receive an answer. It had been reported that Henry and Wilfred Leland bought

1922 Lincoln "Brunn" town car. (Source: Leland Collection)

1922 Lincoln phaeton.
(Source: J.A. Conde)

back the original Lincoln Motor Company stock certificates with their own funds, and thus lost their entire fortunes in the venture, totaling approximately $5,000,000. The original Lincoln Motor Company stock certificates are now in the possession of the National Automotive History Collection of the Detroit Public Library.

Three months after the receiver sale of the Lincoln Motor Company, an overzealous congressman from Michigan, along with other politicians, threatened to impeach U.S. Attorney General Harry M. Daugherty for his alleged failure to prosecute the case of the government against the Lincoln Motor Company. Being under political pressure, it was announced on April 13, 1922, that Attorney General Daugherty would personally investigate the war contracts with Lincoln Motor Company. On April 18, 1922, the United States Government filed suit against the bankrupt Lincoln Motor Company in the Federal Court in Detroit, Michigan. The government claimed that the amount of $9,188,561 asked in the suit was the sum of money Lincoln Motor Company was to have been overpaid for the Liberty aircraft engines

After months of investigation, the Justice Department concluded that the government was not overcharged, nor were there any wrongful acts on the part of Lincoln Motor Company, and that Lincoln fulfilled all the provisions and all the obligations of the contracts. This was of very little comfort to the Lelands, after the Lincoln Motor Company was sold prematurely, and perhaps unnecessarily. Henry and Wilfred stayed on with Lincoln until June 17, 1922, when disagreements with Henry Ford resulted in their being forced out of the company.

After leaving Lincoln, Henry and Wilfred Leland remained active, getting controlling interest in the Inyo Chemical Company in California. They also maintained a financial business in Detroit, located at 2230 Dime Bank Building. Henry was very healthy and agile, in fact he climbed 433 steps to the office in the Dime Bank Building on February 16, 1923, his 80th birthday. Henry M. Leland lived very actively until his passing on March 26, 1932, at age 89.

Left to right: Henry Leland, Mrs. and Wilfred Leland, Eleanor and Edsel Ford, Clara and Henry Ford. (Source: Leland Collection)

John

Willys

During the "Roaring Twenties," when most car manufacturers were building big "dinosaurs,"thinking the road to success was "Bigger is Better," John Willys had the foresight to recognize the need for a smaller, economical, and reliable vehicle. It was Willys-Overland that put Toledo, Ohio, on the map. The United States Army also recognized a good thing when it saw it, mandating that the "GP" (Jeep) vehicle be powered by the reliable Willys engine. The rest is history.

* * *

A Salesman from the Start

John North Willys was born on October 23, 1873, in Canandaigua, New York, the only son of David Smith and Lydia North Willys. His father, a manufacturer of tile and brick, wanted John to get a college education and become a professional; however, from his earliest childhood days John enjoyed promoting deals with his friends. He always seemed to have something in his possession he could offer for sale. John engaged in his first venture when he was not quite 14 years old. His product, which he no difficulty selling, was a clamp installed on the "dashboard" of a horse-drawn vehicle to support the reins and keep them from dropping under the horses' feet. Needless to say he was successful, realizing a profit of 100% on each sale.

Before he was 16 years old, John was able to convince his father that it was a good idea to advance him enough money to enable him, with a friend two years older, to purchase a laundry business at Seneca Falls, New York. The young men realized that their business was a tough one, but by working hard and applying shrewd business practice they were able to sell the business a year later to recover their investment along with a profit of $100 each.

John Willys went back to Canandaigua in 1889 and resumed his studies. He finished high school and started working in a lawyer's office with the intent of going to college, but those plans fell through when his father died in 1891, and at age 18 he was again a businessman.

Willys was able to get an agency for the "New Mail" bicycle, and established a store and repair shop in Canandaigua. He prospered at first but the depression of 1893 took its toll on all businesses. Many of his customers could not pay up their old charge accounts, thus he lost not only the money due him, he lost the customers as well, forcing him to give up his business.

His next venture was as a salesman for the Boston Woven Hose and Rubber Company, enabling him to save about $500. On December 1, 1897, John Willys married Isabel Van Wie of Canandaigua. The following year, with the money he had saved, he bought the Elmira Arms Company, a sporting goods store in Elmira, New York. The price was right, because the former proprietor got "Klondike Gold Fever" and was anxious to sell. The Elmira Arms Company handled Remington firearms and ammunition, but Willys added other merchandise and began to specialize in bicycles. The sale of bicycles was so profitable that Willys progressed from retail to wholesale selling. By the time he reached the age of 27, Willys had built the volume of business to over $500,000 annually.

Catching the Automobile Bug

Willys was always looking for new deals. In 1899 he saw his first automobile, a Winton, while on a business trip to Cleveland, and became very interested in automobiles. At the time he was selling bicycles manufactured by the George N. Pierce Company of Buffalo, New York. As soon as Pierce began making the Pierce Motorette, a small automobile, Willys went to Buffalo to investigate the possibility of becoming a dealer. He succeeded, and in 1901, his first year as an automobile dealer, he sold two cars.

The following year he added the Rambler line, and sold four cars, and in 1903 he sold 20. The public demand for cars was increasing so rapidly that the manufacturers were unable to satisfy the demand using the primitive production methods of the time.

American Motor Car Sales Company

In 1906 Willys sold the American motor car, an expensive, powerful automobile manufactured in Indianapolis. The prestigious American sold for more than $3,000, so it was expected that sales would be limited. Thus, Willys was looking for a less-expensive car that offered greater selling opportunities. The Overland car built in Indianapolis met this prerequisite.

The first Overland car had been built by Claude E. Cox at Terre Haute, Indiana, in 1903. The Overland Automobile Company was incorporated on February 19, 1907, in Indianapolis, Indiana, with Claude E. Cox and David M. Parry as incorporators and directors. David M. Parry was the president of the Parry Manufacturing Company, one of the foremost builders of carriages and other horse-drawn vehicles. (See the Parry chapter in this volume and the Cox chapter in Volume 4.)

While Willys enjoyed selling cars at retail, he was intrigued by the idea of a large wholesale selling organization. In 1906, with E.B. Campbell, a successful lumberman from Wellsboro, Pennsylvania, as a partner, Willys organized the American Motor Car Sales Company with headquarters in Indianapolis, with the object of wholesaling the entire output of the Overland and Marion factories, both located in Indianapolis. For 1907

1905 Overland car near the Indianapolis plant. (Source: Willys-Overland Co.)

1907 Overland runabout.
(Source: American Motors
Historical)

the American Motor Car Sales Company contracted for the entire 1907 production of Overland and Marion cars. Since Willys sold the entire output of 47 Overland cars in 1906, he covered his 1907 Overland contract with a $10,000 cash deposit to assure delivery of the Overland cars.

While the buying response to the Overland car was phenomenal, the production was not, so that by the time of the Wall Street panic on October 22, 1907, the Overland liabilities exceeded $80,000. With Parry unable to furnish any more operating funds, Overland production came to a halt.

In Elmira, New York, Willys was extremely anxious. He had advanced Overland $10,000 to bind his car orders, but no cars were being shipped and he received no reply to his inquiries. Finally, on November 29, 1907, Willys boarded a train for Indianapolis. Arriving the next day on Saturday, he went directly to the Overland factory and sought out Claude E. Cox. Cox's report was most discouraging.

Willys Rescues Overland

Overland Automobile Company was in deep financial trouble. It could not deliver the cars contracted for, but worse, it had liabilities of approximately $80,000 which would have to be paid. There was not enough money in the bank to cover the payroll checks. Overland was to face receivership on Monday, December 2, 1907. This affected the business aspects of the American Motor Sales Company, in which Willys invested all of his resources. Willys knew that if Overland went into bankruptcy, he would lose his $10,000 deposit completely. To save his investment, Willys had to find a way to keep the Overland factory building cars. The first step was have the money in the bank by Monday morning to cover the payroll checks. The Overland checking account was $350 short. Since the banks were closed on Saturday afternoon, Willys asked the manager of the Grand Hotel (where he was staying) to cash a personal check. The manager hesitated on the grounds that he didn't have enough ready cash on hand. This is where John Willys put his magnetism and salesmanship into action; he was able to convince the manager of the need to save a potentially important industry and employer for Indianapolis. The manager agreed to use the proceeds of the restaurant and tavern to cash the check. On Monday morning Willys was at the bank with a handbag full of bills and coins to cover the payroll checks.

The creditors and stockholders felt that Willys was their last hope, since the previous management failed "to make good." Since the property of the Overland Automobile Company appeared to be of no value, and a liability rather than an asset, its restoration to a profit-making basis was the only salvation for Willys, his associates in both companies, as well as the creditors.

A meeting with the creditors was called, at which Willys offered them preferred stock at par value, and common stock at 40% of the liabilities. Franklin Vonnegut, an Indianapolis hardware merchant, started the action by accepting the preferred stock offer. Frank Wheeler, president of the Wheeler-Schebler Carburetor Company, was the leader in accepting the common stock alternative.

David M. Parry's creditors let him keep his Overland stock, believing it to be worthless. Willys made an arrangement with Claude E. Cox and David Parry, whereby they would turn over two-thirds of their Overland common stock to E.B. Campbell and Willys, to be divided by them equally. For this Willys promised to save Overland from the financial crisis. Willys was able to convince Campbell to advance him $7,500 for working capital.

Parry and Cox were not to receive payment for the stock turned in. At the final arrangement, most of the common stock surrendered was from Claude E. Cox' original shares. Cox was reluctant, but with Parry's insistence he agreed. Cox was able to retain 66 shares of his original stock.

It was under these circumstances that Willys acquired control and became president of the Overland Automobile Company on January 9, 1908. Claude E. Cox was named vice president and chief engineer. Being a conservative engineer, Cox would have preferred that the Overland Automobile Company move more slowly. He resigned in early February 1908 with great bitterness, giving his reason for resignation as "the differences of opinion between himself and John N. Willys in the management of the Overland Automobile Company." He had no admiration for Willys' drive, enthusiasm, exuberance, and charming personality. He ended all contact with Overland Automobile Company personnel, except for Edward B. Mull, and Miss Ruth Hoober, his former secretary.

Cox went south on vacation and upon his return to Indianapolis in the spring of 1908, he was besieged by Willys' eagerness to purchase the remainder of Cox's Overland stock. Again, Cox was reluctant; however, on May 13, 1908, Cox entered into an agreement with Thomas P.C. Forbes of New York City, representing Willys' interests, for the purchase of the balance of Cox's Overland stock. The negotiations were so slow that Willys became very impatient, and on August 3, 1908, he sent Cox a very caustic letter. Finally, Cox accepted Willys' offer and surrendered the balance of his stock to Forbes. Cox continued an illustrious automotive career in Muncie, Indiana, Minneapolis, and Detroit. The purchase price for Cox's 66 shares of Overland stock was $1,000. Parry held on to his stock until mid-1909, and reportedly received $250,000 for his holdings.

The Overland Automobile Company was saved, and within a year Overland was operating profitably. Willys was able to convince the employees to come back to work by paying them the wages already due them. He inspired all the employees to do their best to save the company. Under Willys' direction the plant was reorganized, and the first year accounted for the manufacture of 465 cars. The sale of these cars established Overland's financial stability. The demand for Overland cars was so much greater than could be built in the available buildings, that temporary carnival tents were put up to provide cover for assembling cars.

By this time the Overland Automobile Company had been reorganized, with Campbell and Parry being elected to the board of directors. A profit of over $1,000,000 was earned during the first year, and everyone at Overland was pleased with that success.

In early 1909, Willys' next step was to purchase and reorganize the Marion Motor Car Company. While his original plans were to provide additional manufacturing for Overland at the Marion plant, he continued to build Marion cars in Indianapolis and sell them through his organization. The first season that the two plants were under Willys' management, his plan was to build 1,500 cars, but actually 4,075 cars were built and sold in 1909. In 1912 Willys sold his interest in the Marion Motor Car Company to James I. Handley. The

1908 Overland Model 24.
(Source: Willys-Overland Co.)

1909 Overland tourist, with a Stoddard-Dayton at left. Willys is driver of the car at right. (Source: Willys-Overland Co.)

Marion car became the Marion-Handley, built and warranted by the Mutual Motors Company of Jackson, Michigan, and merchandised through the New York Motor Sales Company of Troy, New York.

As the result of the rapidly growing popularity of the Overland, Willys desperately needed larger manufacturing facilities. He originally intended to build a new plant in Indianapolis, and obtained an option on 30 acres of land near the city. But in April 1909, Willys heard that the Pope-Toledo plant was for sale.

1909 Overland Model 34 toy tonneau. (Source: Willys-Overland Co.)

Albert A. Pope, in his far-reaching bicycle and automobile empire, had over-extended himself and had to dispose of the plant, which was now idle in Toledo, Ohio. (See the Pope chapter in this volume.)

Willys heard of this news on Thursday, April 1, 1909, just as he was preparing to board the train bound for New York City. He did not think much about the news until the train had started. The more he thought about it, the more the idea appealed to him. He arranged with the conductor and the Pullman porter to change from a New York sleeping car to a Toledo sleeping car, which happened to be on the same train. Willys arrived in Toledo the morning of April 2nd, and with his energetic drive, he visited the factory during the forenoon, and that afternoon was on the "Twentieth Century" Limited enroute to New York. The following morning, Saturday, April 3rd, Willys met with Albert and George Pope (Albert's cousin), and obtained an option on the Toledo property on April 5, 1909, by a deposit of $25,000.

The reported selling price was quoted at between $350,000 and $400,000. Only 48 hours elapsed between the start of negotiations and obtaining the option. The purchase of the Toledo plant gave Willys the production facilities he needed to keep pace with increased sales of the Overland car. Not only did the Toledo plant provide approximately 3,000,000 sq. ft. of manufacturing area, it was also one of the newest and best-equipped plants in the United States at the time.

Willys-Overland is Born

On October 26, 1909, the Willys-Overland Company was incorporated in Toledo, under the statutes of the state of Ohio, capitalized at $2,000,000. This brought all the Willys plants and companies under the control of the Willys-Overland Company. John N. Willys was elected president, and David M. Parry and Thomas P.C. Forbes were elected to the board of directors.

The year 1909 was certainly successful for Willys-Overland; they built and sold both four- and six-cylinder-engine-powered cars, for a total of 4,860 units. And 1910 was even better, selling over 15,000 cars, all of

1910 Overland. (Source: Willys-Overland Co.)

1910 Overland Model 38. (Source: Willys-Overland Co.)

1910 Overland Model 41. (Source: Willys-Overland Co.)

which were four-cylinder models. Willys could have sold more Overland cars, but was again limited by production capabilities. On July 28, 1910, Willys-Overland capitalization was increased to $6,000,000.

It may be remembered that Willys and E.B. Campbell received two-thirds of the Overland common stock surrendered by Cox and Parry at the time Willys obtained control of Overland in January 1908. Thomas P.C. Forbes purchased the balance of Cox's holdings, 66 shares of Overland common stock in August 1908. Henry F. Campbell (son of E.B. Campbell) had also invested in Overland stock.

In early 1910, Henry Campbell and Forbes perceived that John Willys felt that they were hampering Willys' management of Willys-Overland. In April 1910, Campbell and Forbes offered to sell their holdings of Overland stock to Willys for $2,000,000, but Willys offered only $450,000. Campbell and Forbes took the matter to court, in a suit against John N. Willys and American Motor Car Sales Company. A compromise was reached on June 1, 1910, in the Indianapolis Court. While the terms of the settlement were not disclosed, a sizeable amount must have been awarded, because after the departure of the Campbells from Willys-Overland, Henry F. Campbell was able to finance Harry C. Stutz' Stutz Auto Parts Company, and the subsequent Ideal Motor Car Company, which later merged into the Stutz Motor Car Company.

Over the next couple of years Overland proliferated in the number of models available, to the point that company personnel referred to them by numbers (58, 59, 60, 61) for identity. The 1912 Overland models featured a new "all-steel," four-door, touring car, with the body built by the Edward G. Budd Manufacturing Company. The production and sales of Overland cars kept increasing to over 26,000 cars in 1912, so the capital stock of Overland was increased to $25,000,000 for expansion to keep pace with the sales.

New Acquisitions

Back on April 1, 1905, Arthur L. Garford, president of the Garford Company, had purchased the assets, business, and Elyria factory of the Federal Manufacturing Company to manufacture automotive axles, transmissions, steering gears, hubs, carburetors, mufflers, steel dashes, and sheet metal parts. It was incorporated in the state of Ohio with a capital of $400,000. Touring car parts were made and sold separately, assembled into a complete chassis. Garford later entered into a contractual agreement with the Studebaker Brothers, whereby Garford would build the complete chassis for the Studebaker bodies and coachwork. These cars were merchandised by the Studebaker sales organization as the Studebaker-Garford 40.

In 1908, when the Everitt-Metzger-Flanders Company was organized and in operation, Studebaker entered into a contractual arrangement under which Studebaker obtained exclusive rights to sell the entire output of E-M-F products known as E-M-F Studebaker cars. Thus, when Studebaker acquired control of E-M-F in March 1910, and merged into the Studebaker Corporation on January 1, 1911, it was decided to build their own Studebaker 40.

Garford, after parting from Studebaker, continued to manufacture automotive components and Garford chassis for the Berg, Cleveland, and Rainer cars, on which custom bodies were built. Garford also built complete cars which featured both four- and six-cylinder engines. The Garford 60 Town Car at $4,500 was an exclusive car for the wealthy. Garford also built trucks which were merchandised along with the Garford cars, since July 1911, by the Willys-Garford Sales Company, a subsidiary of the Willys-Overland Company. On July 10, 1911, Willys bought the entire assets of the Garford Motor Company of Elyria, Ohio, for $2,800,000 and took over the plant on August 1, 1911.

On March 26, 1912, Willys purchased the Gramm Motor Truck Company of Lima, Ohio. Garford cars and trucks as well as Gramm trucks were later merchandised by the Willys-Overland sales organization.

In the meantime, C.G. Stoddard and H.J. Edwards, officers of the United States Motor Company, resigned their positions

A.L. Garford

1912 Garford G-8. (Source: Willys-Overland Co.)

1912 Garford G-14. (Source: Willys-Overland Co.)

and organized Edwards Motor Car Company on February 12, 1912, incorporated under the laws of the state of Maine. The Edwards Motor Car Company headquarters were at 1790 Broadway, New York City, and the factory was in Long Island City. Edwards and Stoddard were able to retrieve the Knight engine license rights given up by Stoddard-Dayton with the demise of United States Motor Company.

The Edwards-Knight car was introduced in December 1912, and featured a four-cylinder Knight sleeve-valve engine having a bore of 4 in. and a 5.5-in. stroke. The 120-in. wheelbase chassis had a full-floating rear axle,

EDWARDS-KNIGHT

"The Achievement of Experience"

Designed by H. J. Edwards—one of the leading engineers of the automobile world with a record of ten years' successful accomplishment.

The elimination of experiment — Not a single untried unit in the entire construction. Every feature tested by years of use. The Knight Engine—Worm Drive Rear Axle—Four Speed Transmission, Direct on Third—Lanchester Spring Suspension—Wire Wheels—S. U. Carburetor and American Simms Magneto—Oil flow automatically increased as the throttle opens—these are some of the tried and proven features of the Edwards-Knight.

The elimination of expensive upkeep — A natural and safe ratio of power to weight is essential to reliability, sturdiness and economy. When the highest standard of materials and workmanship are combined with common sense in design a ratio of one-horse power per 100 lbs. weight is always safe—the ratio of the Edwards-Knight.

Economy in Gasoline Consumption — High priced gasoline compels attention. Comparatively small power units economize gasoline. A low gear ratio with direct on third makes a snappy car with high speed possible on fourth gear.

The elimination of exorbitant tire bills — That Wire Wheels add life to tires has been demonstrated by exhaustive tests in Europe, where they are universally used, and, when combined with the right Spring Suspension, maximum tire life is assured.

Silence — The Knight Engine, admittedly the most silent of all gasoline engines, combined with a Worm Drive Rear Axle, results in an elimination of noise which is only equalled by electrics.

Comfort — Silence, combined with a Spring Suspension which precludes the possibility of sudden jolts and jars, or the need of shock absorbers, means comfort. The Edwards-Knight is the easiest riding car in the world—yes, we mean just that and can prove it.

These points, combined with graceful body designs, elegant fittings and other features, make the Edwards - Knight the easiest selling car on the market.

Touring Car, $3500; 4 - Passenger Torpedo, $3500
Speedster, $3500; Roadster, $3500; Limousine, $4600

Every Car fully equipped

Agents who follow closely the trend of the times, who are quick to take advantage of changing tendencies to increase their sales, will be interested in the Edwards-Knight proposition. Write today for full particulars.

The EDWARDS MOTOR CAR CO., 1790 Broadway, New York (Entire Fourth Floor)

1914 Willys-Knight limousine. (Source: Willys-Overland Co.)

worm drive, Lanchester cantilever rear springs, a USL dynamo-flywheel-type electric starting and charging system, a unit multiple dry-disc clutch, and a four-speed gearset, in which the fourth speed was overdrive. Six body styles were offered in attractive styling, including wire-spoke wheels as standard equipment. While the Edwards-Knight was a handsome vehicle, the buying customers stayed away due to the unattractive pricing of $3,500 to $4,700 and the lack of dealer representation. Thus, when John Willys offered to buy the company, Edwards and Stoddard jumped at the opportunity to sell. On October 29, 1913, Willys bought the Edwards Motor Car Company, along with the license rights to the Knight sleeve-valve engine.

Willys had all the tooling, machines, components, partial assemblies, and technical documentation moved to the Elyria plant he had purchased from Garford. For 1914 the car was renamed Willys-Knight, and was basically the Edwards-Knight with a Willys badge, with the same expensive price tag.

On October 12, 1913, Willys purchased the Globe Ball Bearing Company of Norwich, Connecticut. All the machinery and equipment was moved to Elmira, New York, where it was re-established under the Morrow Manufacturing Company, a Willys-Overland subsidiary. The net earnings of the Willys-Overland Company for the year ending June 30, 1913, were $5,705,537.

The Willys-Knight was manufactured in the former Garford plant in Elyria during 1914. However, with the Overland sales increasing during 1914, Willys-Overland enlarged the Toledo plants to 64 buildings having a floor area of 2,623,600 sq. ft. (about 60 acres). It was reported at that time that it was perhaps the largest manufacturing plant in the world.

Showing Appreciation

John N. Willys was always a compassionate and benevolent employer, providing such facilities as a country club, tennis courts, boat house, and other recreational activities for the benefit of his employees. He enacted "Overland Day" on June 23, 1914, and entertained his 10,000 employees with a professional baseball exhibition game of the leading American and National League teams, the Philadelphia Athletics and the Chicago Cubs. Each team received $5,000 and $500 was paid for the use of the park. All of the 10,000 employees marched from the factory to Swayne Field, led by Willys, and received a full day's pay. Both teams played their regular line-up, and the final score was Cubs 8, Athletics 7. Willys personally escorted Connie Mack, famous Athletics manager, through the Willys-Overland plant, and it was one of the most beneficial acts ever performed by an American manufacturer.

Still Growing

The demand for and sales of the Willys-Knight increased during 1915 due to the reduced price, possible by the more efficient production facilities. Both the Overland and Willys-Knight featured a switch box on the steering column for lights and ignition switch with a lock. The 1915 models consisted of: the Willys-Knight, Overland Model 80, Model 81 priced at $850, and the Overland Six Model 86 priced at $1,475. All Willys-Knight and Overland models featured two-unit starting and charging systems, semi-elliptical springs in front suspension and three-quarter springs in the rear suspension, and the standard transmission-axle in the rear. The styling was streamlined, with concealed door hinges. The assembly of the Willys-Knight was moved to the Toledo plant, but the Willys-Knight engine continued to be built in the Elyria, Ohio, plant.

On January 18, 1915, Willys sold his interest in the Gramm and Garford companies to a bankers group, the Geiger-Jones Company of Akron, Ohio. The group merged truck units into the Gramm production facilities at Lima, Ohio, but each truck line retained its own identity. By contractual agreement Willys continued to sell both Garford and Gramm trucks through the Willys-Overland sales organization through 1918. In 1915 he bought the controlling interest in the Electric Autolite Company of Toledo, Ohio, Warner Gear Company of Muncie, Indiana, New Process Gear Company of Syracuse, New York, and the Fisk Tire and Rubber Company of Chicopee Falls, Massachusetts, all under Willys Corporation control.

On October 1, 1915, at the recommendation of the New York banking firms, Willys engaged Clarence A. Earl as vice president and general manager of Willys-Overland. During 1915 Willys-Overland car production amounted to 91,780 units, and the earnings exceeded $11,000,000. In January 1916 Willys-Overland capitalization was increased to $75,000,000.

1916 Overland 75. (Source: John A. Conde)

The 1916 Willys-Knight and Overland cars were refined and improved. The Overland 86 was continued, and the Model 83 superseded the 80 and 81. A new lower-priced Overland 75 was introduced during 1916. Willys-Overland sales amounted to 142,779 units, and the earnings rose to $10,016,420. During 1916 Ward M. Canaday joined Willys-Overland as advertising manager.

The War Years

On April 6, 1917, the United States declared war on Germany, Austria, and Hungary. Many Willys-Overland plants had to be converted for the production of war materiel, and steel shortages forced the elimination of some of the models, namely the 83 and 86. Car production efforts were concentrated on the Willys-Knight, the improved Overland 75, and the newly introduced Overland 90. Export sales remained strong. The manufacture of aircraft engines and components required large expenditures for plant additions, machines, tools, dies, and other equipment, yet Willys-Overland earned a profit of $8,500,000 for the twelve months ending on December 31, 1917.

At this time the design and engineering departments were very active in the development of a new and smaller Overland "Four" in the $500 price range to compete with Ford. Announced on October 4, 1917, the Overland Four Model 91 was a very attractive touring car, affectionately called the "Baby Overland," with a wheelbase of 100 in. But, with a perimeter frame in the shape of an ellipse, with the unique V-form placement of the front and rear quarter-elliptical springs, it in effect gave a riding quality equal to cars of a longer wheelbase. The Overland Four weighed 1,500 lb., and was powered by an L-head four-cylinder engine having a bore and stroke of 3.25 x 4 in. The Overland Four was formally introduced at the National Automobile Show in New York in January 1918, but did not get into production until October 1919, due to the war and the strike against Willys-Overland in 1919.

On June 18, 1918, Willys-Overland made all manufacturing facilities in Toledo, Elyria, and Elmira available to the United States Aircraft Production Board. Willys bought the controlling interest in the Curtiss Aeroplane and Motors Corporation of Buffalo, New York, and the Willys-Overland and Curtiss plants built approximately 1,000 Sunbeam aircraft engines for Great Britain, and 8,412 Curtiss OX-5 for United States aircraft. Willys-Overland was the largest supplier of aircraft engines to the United States Navy. They also

1917 Willys-Knight.
(Source: American Motors Historical)

1918 Overland 90. (Source: Willys-Overland Co.)

produced a great quantity of munitions and military vehicles. Even with the production of cars being limited, and the profit on war materiel controlled, Willys-Overland earned a profit of $11,515,645 in 1918.

On September 6, 1918, Willys bought the controlling interest (approximately 82%) of the common stock of the Moline Plow Company, with the intent to sell farm implements and parts.

After the Armistice on November 11, 1918, the permanent offices of the Willys Corporation were established in the Vanderbilt-Concourse Building in New York City on April 29, 1919. John Willys expected to spend most of his time there, as the pressure of war production and the supervision of the scattered subsidiary plants, required that he spend a lot of time away from Toledo, Ohio. During his absence, vice president Clarence A. Earl was put in charge as general manager of the Toledo plant.

The Strike

In 1919 everyone was making the transition from war efforts to peacetime living and production. The returning military men, labor unrest, and a downturn of the economy all strained labor relations and caused confusion and distrust. All cities were affected and Toledo was no exception. On January 27, 1919, John Willys, in a meeting with the plant foremen, announced a profit-sharing plan whereby Willys-Overland would divide the profits equally between capital and labor, after wages and profit in relation to the investment were taken out.

The profit-sharing plan was retroactive to January 1, 1919, and the wages in effect from time to time would not be affected. It was understood that every employee would benefit, and these plans would have no effect on periodic wage adjustments. The plan would recognize and reward individual and department efficiency, and increase the reward for continuous service. The first distribution under the profit-sharing plan was made on April 21, 1919, with $200,000, and was made to employees who had been with the Willys-Overland Company for more than six months.

Even though Willys was admired and well liked by most of his employees, and despite the profit-sharing plan, on March 27, 1919, the Willys-Overland Company was presented with written demands by a committee of employees claiming to represent all employees. The demands, among other considerations, called for increases in pay from 15% to 50%, a 44-hour week of 5 days at 8 hours and 4 hours on Saturday, the elimination of all piece work, and double-time pay for all overtime. The demands also called for a "closed shop" whereby only employees represented by the union would be permitted to work, and that a committee elected outside of the company's plant would be recognized by the company for the adjustment of all grievances.

Since John Willys was not in Toledo at the time the union demands were made, it was up to Clarence Earl to handle the labor relations problem. Earl did not possess the diplomacy, personality, and salesmanship of Willys. Hence, when Earl rejected the union demands on April 17, 1919, the employees threatened a future walkout. The company learned that on Saturday night, May 3rd, the union had instructed its members to quit work at 3:30 on Monday, May 5th. On Sunday May 4th, the company had posted notices on all bulletin boards throughout the plants outlining the situation and stating that the men who left their job at 3:30 p.m. did so against company rules and would be fired.

On Monday, May 5, 1919, the International Machinists Union called a strike at the Willys-Overland Company plant at 3:30 p.m. Approximately 6,000 workers quit at that hour, refusing to work until 4:00 p.m., the normal closing hour. The following day about 6,665 out of 13,000 men reported for work. The strikers picketed the factory and declared they would exert every effort to force the company to meet their demands. The automobile production was considerably curtailed.

Conferences were held all week regarding the strike and both sides agreed to have the matter resolved by arbitration. On Saturday the Willys-Overland Company took out a full-page advertisement in the *Toledo Blade* which stated the company's position in the labor trouble. It quoted that only approximately 1,200 men were present at the meeting when the strike vote was taken, and that the decision to strike affected all employees. It further stated that "In justice to the company and to the many thousands of loyal employees, the company did not and could not recognize the right of a small group of men to determine the company policy, and to so vitally interfere with the inherent rights of 13,000 employees." The ad also stated that the factory would be open on Monday, May 12th.

Management's attempts to reopen the plant and resume production with non-striking employees was not successful because loyal employees who were willing to work feared for their safety. Work was resumed on May 26th after two weeks of idleness, and those who came in to work were taunted and harassed by striking picketers. Finally, on the night of June 3, 1919, rioting broke out at the Willys-Overland plant as a disorderly mob of several thousand attacked loyal workers as they were leaving the factory. Earlier that day, street cars carrying loyal workers were attacked, windows smashed with bricks, and the passengers beaten. The riots occurred at the entrances and exits of the plant, and at street car transfer points. Many were injured and two men were killed during an attack in the home district of loyal employees.

Mayor Schreiber of Toledo urgently requested the Willys-Overland management to close the plant and he declared that the city was unable to protect the factory and its employees. The mayor appealed to Governor Cox of Ohio for troops to protect the city. The plant area of Toledo was placed under martial law to protect the plant and the workers. Several hundred federal, state, and local peace officers were called in to maintain law and order.

The company was aided by an important court decision which resulted in the issuance of an injunction by Judge John J. Killits, which ordered "the Willys-Overland Company to resume operations at once and to

1920 Baby Overland Four. (Source: Willys-Overland Co.)

continue operating for a period of ten days." The Willys-Overland plant reopened on June 17, 1919, with all departments in operation. The plant was operating under the protection of the United States District Court, and was guarded by 279 Special Deputy Marshals, National Guard soldiers, policemen, and guards.

1920 Baby Overland Four. (Source: Willys-Overland Co.)

The injunction by Judge Killits was made permanent on June 24, 1919, which limited the number of picketers and their behavior. It also provided for the deputizing of additional United States Marshals to enforce the order. Judge Killits' decision was a sweeping one watched by Detroit, and may have set a precedent which other courts could follow in similar strike cases. By July Willys-Overland was operating fairly well, but full production was not achieved until the end of 1919.

Thus the Baby Overland did not get into production until October 1919, two years late. By this time inflation had set in, and the costs of production and material had increased so much that it raised the selling price of the Overland Four to a level no longer competitive with the Ford Model T. The post-WWI depression had set in leaving the Willys-Overland Company with a $14,000,000 debt to suppliers.

Chrysler Steps In

In August 1919 Willys purchased the plant facilities of the Duesenberg Motors Corporation at Elizabeth, New Jersey, where the famed U-shaped King-Bugatti aircraft engine was developed and built for the U.S. Army Air Corps. The Duesenberg brothers moved to Indianapolis, and in the Engineering offices, Carl Breer, Owen Skelton, and Fred Zeder were set up to develop a new light car with a new high-performance six-cylinder engine (more on that later).

On November 10, 1919, Walter P. Chrysler abruptly resigned his position as president of Buick Motor Company and executive vice president of the General Motors Corporation. The reason given was because of his disagreement with William C. Durant over corporate operating practices. Since Chrysler held a considerable number of shares of General Motors stock and was about to dispose of it, the General Motors directors and the banking firms thought it more prudent to have the duPont interests buy the shares from Chrysler rather than have them sold on the open market, which could have made the value of General Motors stock plunge.

1920 Willys-Knight Model 20 touring car. (Source: Willys-Overland Co.)

Meanwhile, John N. Willys was in England trying to arrange a working alliance with the Crossely Motor Works Limited to build a lighter low-priced car. Willys owed the banking firms $18,000,000 and had $14,000,000 of unpaid debt to the suppliers. This made the bankers uneasy about the future of Willys-Overland. Willys, knowing Willys-Overland was in deep financial trouble, again sought financial help from the banking firms.

The bankers were willing to help Willys-Overland again, but this time to protect their financial interests they demanded that Walter P. Chrysler be put in full control of Willys-Overland and the Willys Corporation. Chrysler joined Willys-Overland on November 1, 1919, and on January 6, 1920, he was appointed executive vice president and general manager of the Willys-Overland Company. These facts were reported in the January 10, 1920, issue of *Automotive Topics*, and the January 8, 1920, issue of *Automotive Industries*.

Some of the published accounts of the Willys-Overland struggle tend to credit Walter P. Chrysler alone for the restoration of Willys-Overland to a viable position. What Willys did was not because of Chrysler, but because Willys alone would have done it anyway. Most of the corrective action taken by Willys-Overland occurred after Chrysler's departure from the company. It is true that Chrysler could see Willys-Overland's problems with an unbiased view. Chrysler in effect was the "new broom" that could and did sweep cleanly. In Chrysler's view the problems were lack of stringent cost control, operating control in the hands of incompetent persons, over-extension of plant facilities, diversification into other unrelated fields of business, and what Chrysler thought was an uninteresting and unsaleable product. With these thoughts in mind he proceeded to do something about it.

One of his first moves was to cut John Willys' salary in half, not so much to effect a savings for Willys-Overland but to show everyone who was running the show. Willys took the cut gracefully, and at no time did he oppose Chrysler's actions. Other actions were taken to reduce and control costs. Chrysler replaced those in operating control, including Clarence A. Earl in December 1920. While the news media and public relations releases called it a resignation, Earl must have known it was coming. Many of Earl's subordinates loyal to him had already been dismissed, and Earl had his house listed for sale with a real estate agency. The reason given for Earl's departure was the moving of the Willys-Overland Executive Offices to New York City.

Charles B. Wilson, general manager of the Wilson Foundry and Machine Company of Pontiac, Michigan, a Willys subsidiary, was named general manager of all production facilities of Willys-Overland and Willys Corporation plants. In effect Wilson became Chrysler's right-hand man. W.R. Kilpatrick continued as the production manager of the Toledo plant.

By this point Chrysler was having another automotive "affair," this time with the Maxwell-Chalmers group as the chairman of the management committee. By January 1921 Chrysler was handling the business affairs of Willys-Overland, Willys Corporation, and the reorganized Maxwell Motor Corporation and Chalmers Motor Company now known as the Maxwell-Chalmers Motor Corporation. In 1921 it was referred to as Chrysler's "dexterity"; today it might be called a conflict of interest. On January 1, 1921, *Automobile Topics* headlined its article "Walter P. Chrysler is a very busy man."

Walter P. Chrysler and John N. Willys were headquartered in the Willys Corporation offices in New York. Chrysler remained in New York handling financial affairs, while Willys spent most of his time traveling from city to city, encouraging and inspiring dealers. This not only stimulated the dealers, but Willys himself. After seeing Willys again and being infected by his enthusiasm, there was no question that the dealers were fully behind him.

The Recovery of Willys-Overland

A new distributing policy was put into effect in early January 1921, whereby the Willys-Overland plant in Toledo would produce cars only on actual orders from distributors and dealers. New car inventories at the factory and distributors were reduced to only a two-week supply of cars. This in effect provided more working capital and enabled Willys-Overland to pay off some of the obligations to the banks. By the end of January 1921, Willys-Overland was rehiring some of the laid-off workers in Toledo; 2,500 Willy-Overland workers were back at the Toledo plant by September. Willys-Overland obligations to the banks were reduced to less than $20,000,000, and 400 new dealers were added to the Willys-Overland sales organization by November 1921. Due to aggressive selling by the dealer organization, and with the improved national economy and improved worker efficiency, Willys-Overland was now operating on a sound profitable basis. On September 6, 1921, Willys-Overland cut the prices of their cars: $595 for the Overland touring, and $1,525 for the Willys-Knight touring f.o.b. Toledo. By the end of 1921, Willys-Overland had 6,000 workers on payroll in Toledo, and had contract orders for 70,000 cars for 1922. Walter P. Chrysler left the Willys-Overland Company at the end of 1921.

The Breakup of Willys Corporation

Unfortunately, the over-extended Willys Corporation was losing money, even though some of its divisions were operating profitably. On December 1, 1921, the Willys Corporation board of directors consented to receivership action, independent of the bankers, and started conferences for reorganization. The receivership action did not affect Willys-Overland Company, even though 30% of its stock was owned by the Willys Corporation. The purpose of the receivership was to split up the Willys Corporation into separate units or companies, allowing each to operate independently, and also to conserve the assets and to avert hostile action by disgruntled creditors. The financial embarrassment of the Willys Corporation came about when its operations were overtaken by the general business depression which started more than a year before.

The Elizabeth, New Jersey, plant was a major cause of Willys Corporation's financial difficulties. Shortly after buying the former Duesenberg factory, which was almost new, Willys Corporation awarded a contract to

1921 Willys-Knight roadster.
(Source: Willys-Overland Co.)

the American Concrete Steel Company for the demolition of the former Duesenberg plant and the erection of a new factory building. Most of the work was done on a cost-plus basis, when wages and material costs were at their peak. The Willys Corporation had planned to manufacture the new Chrysler Six at the plant. However, the car introduction planned for the year before was repeatedly delayed and frowned upon by the New York bankers, who were heavy creditors but had little or no interest in building cars. Their main interest was to get their money back by selling the plant, which they felt was a white elephant anyway.

Unlimited development of the Electric Auto-Lite Corporation division was assured under the plan of receivership. The Electric Auto-Lite division had some of the most successful and most profitable plants of their kind in the United States. Even in face of the pending receivership and adjustment, it operated at an enormous profit. The Electric Auto-Lite, U.S. Light and Heat, and the New Process Gear plants had not been closed at any time during the depression, and each had returned a substantial profit.

During 1922 and 1923 Willys and the receivers directed much of their effort to selling off some of the facilities and assets of the Willys Corporation. At a board of directors meeting on April 6, 1922, Willys-Overland declared a net loss of $8,633,279 for 1921, compared to a total deficit of $5,480,394 for 1920. Willys-Overland took these losses and then moved ahead on a profit-making basis in 1922. On April 9, 1922, at a board meeting, consent was obtained to reorganize Moline Plow Company and to scale down the securities to a realistic level. This action was taken to protect Moline from total collapse.

On April 24, 1922, it was announced that the Elizabeth plant would be put up for sale on auction by the receivers on June 9, 1922. C.B. Wilson was elected president of the Willys-Morrow Company at Elmira, New York, and the Wilson Foundry and Machine Company of Pontiac, Michigan, on May 24, 1922. On June 6, 1922, Clement O. Miniger and associates purchased the Electric Auto-Lite division in Toledo, Ohio, at a special "master's sale" for the minimum price of $4,700,000 set by the Federal Court. This sale upset the New York bankers. The New Process Gear plant facilities in Syracuse, New York, were also sold in June 1922.

On June 9, 1922, the huge Elizabeth plant was purchased at the receiver's sale by Durant Motors Incorporated for $5,525,000. Walter P. Chrysler and the Maxwell Motor Corporation were represented by James Smith Jr. and bid up to $5,500,000. Durant Motors' intent was to build the new Star Car in the Elizabeth plant, which was to be introduced in the fall of 1922. Had Chrysler been successful in his bid for the plant, it would have been used for the production of the Chrysler Six, the purpose for which it had been built. The rights to the design of the car passed on to Durant as the result of the sale of the plant.

It must be remembered that the Chrysler Six car was designed by Carl Breer, Owen Skelton, and Fred Zeder while they were in the employ of the Willys Corporation. Just prior to the sale of the Elizabeth plant, Breer, Skelton, and Zeder set up a consulting engineering firm in Newark, New Jersey. Upon the purchase of the plant, Durant took the design drawings of the car (which came with the plant) to Breer, Skelton, and Zeder at Newark. He requested certain changes, including a change to a six-cylinder Continental engine and a midship transmission, to facilitate the purchase of these two units as shelf items which would eliminate the need for him to manufacture them. He also requested that the radiator shell be reshaped to resemble that of one of the most expensive cars in America, the Locomobile. The design finally emerged as the production model of the new Flint for 1924.

The changes requested by Durant left the rest of the Breer, Skelton, Zeder car, including the original high-performance, seven-main-bearing, six-cylinder engine with unit transmission, and the newly designed radiator shell intact to be had for the asking. Chrysler, in 1923, induced Breer, Skelton, and Zeder to come to Detroit and join him to develop and complete the car they originally intended to be the Chrysler Six. In retrospect, those who might remember the 1924 Chrysler and the 1924 Flint will easily recognize the appearance similarities of these touring cars. The most noticeable difference was the shape of the radiator shell and Chrysler's winged radiator cap.

Willys Saves Overland Again with a New Line-up

While Chrysler and Durant were competing in their relative fields, John N. Willys was not idle. All during 1922 and 1923 Willys was on the road most of the time, traveling from city to city, meeting and inspiring dealers. Even the "uninteresting and unsaleable" cars for which Chrysler had such disdain were selling in record numbers. Although the (Baby) Overland Four was not a serious competitive challenge to the Ford Model T because of price, the Overland Four represented much more value to those who knew cars, and wanted more than just transportation. People bought Ford Model Ts because of their simplicity and price.

For 1923 while there were no great mechanical changes in the Overland except minor improvements, the styling was considerably improved by raising the hood and radiator shell about 1.5 in. to provide a straight line from the shell to the cowl. The Willys-Knight introduced a new close-coupled model called a coupe sedan. The other Willys-Knight models were improved and disc wheels were offered as an option.

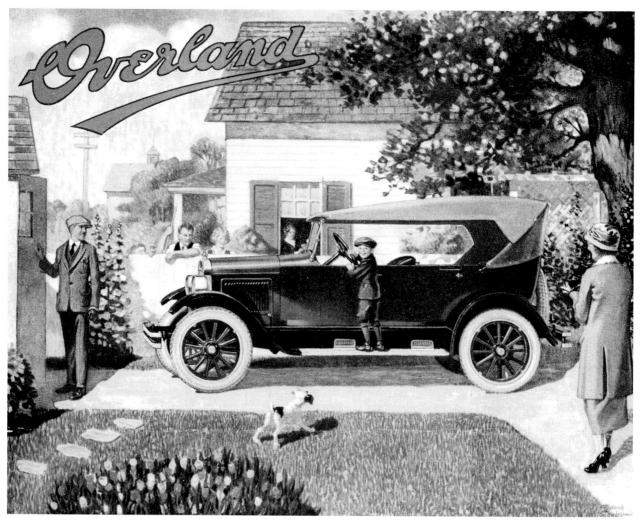

1923 Overland Red Bird. (Source: Willys-Overland Co.)

In May 1923 the Overland Red Bird was introduced, a new touring car painted cardinal red with black fenders and running gear. The upholstery was a matching red, and the folding top was of tan fabric. The radiator shell and the drum-shaped headlamps were nickel-plated. The Red Bird certainly boosted Overland sales. On October 4, 1923, Overland introduced the Champion, an all-purpose, four-door sedan priced at $695 f.o.b. Toledo. The restyled Overland, the improved Willys-Knight, the Red Bird, and the Champion helped Willys-Overland build and sell 196,038 cars in 1923, netting a profit of $13,034,032. Thus, Willys, by dressing up his cars and pulling himself up by his bootstraps, was able to turn the Willys-Overland Company around from an $8,633,279 net loss in 1921 to a net profit of $13,034,032 for 1923.

For 1924 the same model line-up was continued. On March 17, 1924, the Willys Corporation receivers announced that it had paid all debt claims against the Willys Corporation in full by disbursing $15,300,000. It was only the existing U.S. Government claim of $1,500,000 against Duesenberg Motors Corporation that prevented the dissolution of the Willys Corporation receivership. On the same day the price of the Overland Champion sedan was reduced to $655 f.o.b. Toledo.

The new Overland Blue Bird was introduced on May 7, 1924, replacing the Red Bird. The touring car was painted peacock blue with black fenders and running gear. The upholstery was a matching blue, as was the top deck material. The radiator shell, headlamps, windshield stanchions, and door handles were nickel-plated. The Blue Bird featured 31x5.25 balloon tires and disc wheels, the first use of balloon tires on a low-priced car, listing at $725 f.o.b. Toledo. All mechanical specifications were the same as those of the Overland Four for some time, including the "triplex" spring arrangement. A new business coupe priced at $650 f.o.b. Toledo was introduced on June 16, 1924, helping the Overland sales. The bank indebtedness of $4,200,000 was paid off during June, leaving only funded indebtedness, and a net profit of $2,086,645 for 1924. At the end of 1924 Willys-Overland Company listed assets at $33,264,818 and liabilities at $3,774,974.

Willys-Overland broadened its market coverage with the introduction of the new six-cylinder models of the Overland and Willys-Knight on January 1, 1925. The Overland Six was offered in two-door coach and four-door sedan models. The Willys-Knight Six line consisted of roadster, phaeton, coupe, brougham, coupe sedan, and sedan models. The Overland Six featured balloon tires, and the Willys-Knight Six had balloon tires and mechanical four-wheel brakes. The four-cylinder Overland and Willys-Knight were continued without change, except the Willys-Knight Four hood lines and radiator shell had been altered to conform to the appearance of the Willys-Knight Six.

1925 Willys-Knight Model 65. (Source: American Motors Historical)

During March 1925 Willys-Overland set a production record of 24,300 Willys-Knight and Overland cars, with employment of almost 20,000 at the Toledo, Pontiac, and Elyria plants. The first six month earnings (November 1, 1924, through April 30, 1925) amounted to approximately $14,000,000. The Willys-Knight and Overland Six were worthy competitors in the low-medium and medium price ranges. Domestic new car registrations of Willys-Knight and Overland for 1925 amounted to 157,662 new cars. Export and Canadian sales raised the Willys-Overland total to about 220,000.

Willys-Overland Company had a market price spread from the low-priced, medium-priced, up to the upper medium-priced field. But Willys was in search of a high-priced prestige car. The F.B. Stearns Company, producers of the Stearns-Knight, was one of the first automobile companies founded in the United States (organized in 1898). It was conservative in its operations and was not in financial trouble. In fact, the F.B. Stearns Company reported a profit of over $450,000 for 1924, that being on the level of the earnings for the previous four years. Willys' interest in the Stearns car was to give him that prestige car in the upper price range and thus become the world's largest builder of Knight-engined cars. Acquiring the F.B. Stearns Company seemed to be the most logical move to fill out the complete line of cars in every price range. Therefore, on December 15, 1925, Willys and his associates purchased the assets, plant, name, and good will of the F.B. Stearns Company of Cleveland, Ohio, at a reported price of $2,500,000.

For 1926 Willys-Overland reduced the number of body types and planned for the new four-cylinder Overland to supersede the Overland Four. In June 1926 Willys-Overland introduced the new "Overland Whippet," a totally new smaller car, with features found previously in more expensive cars. It featured four semi-elliptical springs, mechanical four-wheel brakes, a new efficient four-cylinder engine with full pressure lubrication, water pump engine cooling, a new driveline with a dry disc clutch, new transmission, Hotchkiss drive with exposed shaft and universal joints, and a new lighter bevel-gear rear axle.

The Whippet styling was totally new, with a European flair. Perhaps the arrangement with Crosseley Motor Works Limited of England influenced the styling and design of the Overland Whippet. Production of the Whippet started on July 1, 1926, and while the car was well received by the buying public, the delay in

1926 Willys-Knight Model 70A. (Source: American Motors Historical)

1927 Whippet coach.
(Source: John A. Conde)

getting started resulted in the sale of only about 50,000 Whippets in 1926. Thus, the total sales for Willys-Overland dropped, and the reported net profit was only $1,819,689 for 1926.

The new Whippet Six and the new Falcon-Knight Six were introduced in January 1927 at the National Auto Show in New York. The Whippet Four was priced at $625 to $755, the Whippet Six at $765 to $925. The Falcon-Knight Six, featuring a seven-main-bearing Knight engine, was priced in the $1,000 range. The Falcon-Knight was built in the Elyria, Ohio, plant; the headquarters of the Falcon-Knight Corporation were located in the Majestic Building in Detroit. John A. Nichols, former vice president of Dodge Brothers Incorporated, was president of Falcon-Knight Motors Corporation, and Russell Huff was chief engineer.

In January 1927, the F.B. Stearns Company introduced a new Stearns-Knight Model G-8, powered by an in-line eight-cylinder Knight sleeve-valve engine, featuring a nine-main-bearing crankshaft, and developing 100 hp. Available in 18 body styles, the Stearns-Knight was priced from $3,450 to $4,950 f.o.b. Cleveland. During 1927 Willys added three new seven-passenger models in the Willys-Knight Great Six line: a seven-passenger touring priced at $2,495, a seven-passenger sedan at $2,850, and a seven-passenger limousine priced at $2,950 f.o.b. Toledo, Ohio.

In September 1927 the Whippet entered the light commercial car field by introducing four commercial body types: a panel delivery, an open express, and two canopy express models, with and without screens. Three roadster-type commercial vehicles were a sample compartment roadster, an open pickup, and a closed-panel type. The prices ranged from $625 to $710 f.o.b. Toledo. The 1927 domestic sales were 184,127 units, generating a profit of about $6,424,002.

In January 1928 Willys introduced a new Willys-Knight Standard Six, priced lower at $1,145 to $1,245. The Willys-Knight Standard Six shared the same basic body as the Falcon-Knight and the same engine as the Willys-Knight Special Six (formerly the 70A). The former Willys-Knight 66A became known as the Willys-Knight Great Six. The Falcon-Knight prices were increased by $100 on all models, bringing the Falcon-Knight prices within $50 of the price of the Willys-Knight Standard Six for the same corresponding body type. By March 31, 1928, Willys lowered the prices on both the Willys-Knight Standard Six and the Falcon-Knight Six by $150 and $100, respectively, bringing both lines of cars into the $995 to $1,095 range. These price adjustments may have been the reason why Willys-Overland in 1928 had the best sales year ever, moving into third place behind Chevrolet and Ford. Willys-Overland showed a net profit of $8,557,399 for the first nine months of 1928. In April 1928 Stearns-Knight introduced a new six-cylinder Knight-engined car priced at $2,495 to $2,945 as a companion model to the Stearns-Knight Eight.

For 1929, more power, stronger frames, and more attractive body lines characterized the Whippet Four and Whippet Six models. On some of the closed models the prices were slightly reduced giving the Whippet a price range of $485 to $760 f.o.b. Toledo. The cast-iron engine pistons were replaced by aluminum-alloy Invarstrut pistons, which gave smoother performance and increased power. The Willys-Knight Standard Six, Special Six, and Great Six remained substantially the same as the 1928 models except for improved styling lines.

Willys Bows Out

In mid-1929, feeling he could leave matters in good hands, John N. Willys decided to retire from active management of the Willys-Overland Company. He sold his common stock shares to C.O. Miniger, Linwood A. Miller, Thomas H. Tracy, H.C. Tillotson, and other Toledo business leaders. At the board of directors meeting where Willys announced his retirement plans, Miller (former vice-president) was elected president of Willys-Overland Company, and John Willys remained as board chairman.

The Sudden Slip Downhill

The Willys-Overland Company was in healthy financial condition, sales were good, and it had a competent management organization. No one could foresee what the next few months had in store. Even though the 1929 Willys-Overland sales were comparatively good, selling only 18% fewer cars than in 1928 (a record year), Willys-Overland slipped to fourth place because of improved Hudson-Essex sales. Then Black Thursday, October 24, 1929, descended on Wall Street. Even those whom the stock market crash did not affect directly, and who were in the market for a new car, held on to their money, and thus brought on a rapid decline in car sales. Operations at the F.B. Stearns Company plant in Cleveland were discontinued on December 20, 1929, and the company was dissolved on December 31, 1929. Willys-Overland profits for 1929 dropped to $4,023,284.

For 1930 Willys-Overland introduced a new Willys Six with a six-cylinder L-head engine. It featured attractive new styling including a new radiator shell, and was priced at $695 to $850 f.o.b. Toledo. The Willys Six replaced the Whippet Six and the Willys-Knight Standard Six. The Whippet Four continued without change, but the Willys-Knight Special Six and the Great Six were significantly improved in an attempt to bolster sales and reduce manufacturing costs. By late January 1930, Whippet Four prices were reduced by $40 to $50, bringing these models into the $475 to $695 price range.

1930 Willys-Knight Great Six coupe. (Source: American Motors Historical)

1930 Willys-Knight Great Six roadster. (Source: American Motors Historical)

In April 1930 Willys-Overland officially announced its entry into the field of eight-cylinder cars with the introduction of the new Willys Eight as a companion model to the Willys Six introduced in January. Its chassis and styling were patterned on the Willys Six design, including the new radiator shell, with a longer wheelbase and longer hood. The deluxe equipment of the Willys Eight included six wire-spoke wheels, with the spares mounted in wells of the front fenders.

The engine was an eight-in-line L-head type, designed by Willys-Overland and built by the Continental Motors Corporation. The engine had a piston displacement of 245 cu. in. and developed 80 hp at 3,200 rpm, and featured an integrally counterbalanced crankshaft, supported on five main bearings with forced-feed lubrication. Torsional vibrations were dampened by use of a Lancaster vibration damper. The Willys Eight featured fingertip control in the center of the steering wheel, the same as the Willys Six and Willys-Knight models.

In February 1930 John N. Willys was called to Washington, D.C., by President Hoover. They discussed business conditions and the state of the nation's economy. At that time very few people knew the real reason for Willys' visit with President Hoover. On March 17, 1930, Willys was sworn in as United States Ambassador to Poland. Willys enjoyed his tenure as ambassador; he was well-liked by the Polish people because he was an excellent example of a good American citizen. His activities in Poland were very productive, and the relationship between Poland and the United States were the best during his tenure. At home the Great Depression was sinking to its lowest level, and the Willys-Overland Company sold only 80,555 cars during 1930, losing $7,588,392.

For 1931 the styling and designs were refined. The Willys-Knight Great Six 66B had rectangular doors on the hood side for ventilation, clamshell-style front fenders, with wells for the spare wheels and tires. The Willys Six and Eight had horizontal louvres in the hood sides. The Whippet models were discontinued for 1931; all other 1931 models were carried over into mid-1932.

Willys Takes the Reins Again

In 1932 President Hoover felt Willys was needed back in the United States. On April 26, 1932, John N. Willys resigned as the Ambassador to Poland and returned aboard the S.S. Europa on June 13, 1932. When he landed in New York, he was wearing the "Rosette Polonia Restituta" on the lapel of his coat, with a grand cord attached. This was the highest of all Polish Awards, and was given to him in appreciation of his service to their country as United States Ambassador.

Assuming active management of Willys-Overland shortly after his arrival, Willys, in his interviews with reporters, said, "We must get down to fundamentals and work our way out of our difficulties." He also said, "Anyone who has been in Europe, where there has been a depression for 15 years, realizes that what the United States has gone through recently looks like prosperity to Europeans. It's time we stopped being cry-babies and showed more gumption."

Willys turned his attention first to the problems of sales, production, and model line-up; the financial problems he handled later on. By mid-1932, Willys-Overland came out with a new series of cars featuring

1932 Willys-Knight 66D sedan. (Source: American Motors Historical)

streamlined bodies in a modern mode, and equipped with the latest driving conveniences such as: synchronized transmission gears with silent second, free-wheeling, Startix starter control, rubber engine mountings, ride control selector, and many other features. These new cars were the advanced 1932 models which carried on into 1933. The Willys line was offered in three series: Willys Six (Model 6-90A), Willys Eight (Model 8-88A), and the Willys-Knight (Model 66E). The price range was $535 to $1,420.

The advanced 1932 line had a distinctively new appearance, from the gently sloping V front radiator shell, rectangular ventilating doors in the hood sides, sloping windshield, double cowl ventilators, clamshell-type front fenders, to the graceful well-groomed rear quarters and back. Wire-spoke wheels were standard on all models, with demountable varnished wood-spoke wheels optional on the Willys-Knight, and spare wheels and tires carried in the front fender wells. All models featured the convenient fingertip control for starting, lights, and horns, located in the center of the steering wheel. The instrument panel carried the aircraft-style, round-dial, pointer-type instruments which included speedometer, oil pressure gauge, fuel gauge, ammeter, and temperature gauge. This type of instrumentation enabled the driver to observe the indicators at all times with a mere glance of the eye. Despite all the effort that was put into the manufacture and merchandising of these fine cars, the effect of the Depression was so profound that only 25,898 Willys and Willys-Knight cars were sold in the United States in 1932. Export sales were insignificant, since they amounted to less than 1,000 units.

Even before Willys' return, all effort was put forth toward the design of a new low-priced vehicle that would fit into the distressed economy of America at that time, the Willys 77, introduced in January 1933. It was priced at $395 to $475 and featured the same engine as the Whippet Four. It had Floating Power (licensed by Chrysler Corporation) engine mountings, which made the flow of power very smooth. This small car on a 100-in. wheelbase somewhat resembled the British Austin. While the styling may not have appealed to some car buyers, Willys stressed performance and economy of operation in its advertising. It boasted 25 to 30 mpg, when gasoline was selling at 8.25 to 10 cents per gallon.

Tough Times

In 1933 all other Willys cars were discontinued, including the Willys-Knight. While the decision to concentrate on the small economy car was perhaps a good one at the time, good decisions don't always win ball games. There was not only the Depression to contend with, but also the crunch on operating capital. Willys-Overland was in good financial condition when Willys retired from active management in 1929, and the management was in the able hands of Lynwood Miller; however, the financial drain on Willys-Overland, due to enormous losses sustained from 1929 through 1932 (reported at $35,000,000), depleted the operating capital.

Willys reportedly advanced the Willys-Overland Company $2,000,000 of his funds to keep the company going. Since Toledo was hit even harder than Detroit by the Depression, Toledo banks were in no position to loan any money to Willys-Overland. Therefore, Willys went to Detroit to negotiate a loan to bolster the Willys-Overland working capital. By February 13, 1933, he was successful in getting one of the banks to agree to provide the loan. However, at 1:32 a.m. on February 14th, Governor William A. Comstock of Michigan declared a Michigan Bank Holiday, closing all banks for a period of seven days through February 21, 1933. On February 28th, Governor Comstock extended the bank holiday through March 9, 1933. But, during the evening of March 5th, President Franklin D. Roosevelt proclaimed the National Bank Holiday to be in effect for four days only. However, on March 9, 1933, the National Bank Holiday was extended indefinitely, and Congress enacted the Emergency Banking Law.

On February 15, 1933, the Willys-Overland Company went into receivership, and during receivership was permitted to manufacture cars only in small production runs of about 2,000 cars at a time, provided that they

1935 Willys 77. (Source: John A. Conde)

had orders for them. Shortly after receivership, Miller left the company and was succeeded by Ward M. Canaday, former vice president of sales and advertising. Willys-Overland struggled through those hectic days of 1933 and 1934, selling 15,167 cars in 1933 and only 6,576 cars in 1934. Finally, with the help of preferred stockholders and Toledo financiers, Willys-Overland came out of receivership and was reorganized in January 1935. John N. Willys was again elected president, but this presidency was short-lived.

On May 4, 1935, while attending the Kentucky Derby, Willys was stricken with a massive heart attack. With the excellent medical treatment in the Louisville hospital, and with Willys' fighting spirit, he recovered and was home by June 6th. After moving to New York, Willys slowly continued toward recovery, but in August he had another massive attack, and he died on August 26, 1935.

The Company Lives On

On September 5, 1936, the Willys-Overland Company was again reorganized, and the name was changed to Willys-Overland Incorporated. David R. Wilson was elected president and Ward M. Canaday was elevated to board chairman. The Willys Model 77 was continued through 1936, but in January 1937 the 77 gave way to the new Model 37 which was introduced at the National Automobile Show in New York.

The new Willys 37 retained the reliable Willys 77 engine, chassis, and other lightweight features, but was completely restyled in the American motif. It featured an egg-shaped combination radiator grille and hood. The hood and grille assembly opened from the front (alligator style) and had horizontal stamped louvres extending from the center of the grille around the hood sides. The body lines were attractive, similar to the competitive cars of the period, having pontoon-shaped fenders and functional stamped steel wheels.

The Willys 37 was well-received in the United States, with 51,202 domestic sales in 1937. However, the great business slump of 1938, which caused the automotive industry sales to drop from 3,483,752 cars in

1938 Willys 37 after winning Gilmore economy run. Dr. Wilson is at center. (Source: American Motors Historical)

1937 to 1,891,021 cars in 1938, dealt Willys-Overland a severe blow, bringing their sales down to a dismal 13,012 units. The 1937 Willys 37 was a favorite car of the young, and even today many 1937 Willys coupe bodies are used as the shell to house the powerful engines in Drag Racing.

On November 12, 1938, Willys-Overland introduced a second line of cars, identified as the "Willys-Overland." It featured a new X-K frame, and the wheelbase was 2 in. longer. The styling included a distinctive chromium-plated grille, and the headlamps were centered on the front of the fenders giving an "eyeball" effect. The Willys 37 and the Willys-Overland were continued through 1941.

On January 19, 1939, Joseph W. Frazer, who had been with the Chrysler Corporation since 1925 and was vice president of sales for many years, was elected president and general manager of Willys-Overland, succeeding the retiring David R. Wilson. The name of the corporation was changed to Willys Motors Incorporated. Frazer appointed D.G. Roos chief engineer. On April 1, 1939, Willys Motors Incorporated added a new line

1939 Willys 38. (Source: American Motors Historical)

Joseph W. Frazer. (Source: American Motors Historical)

known as the "Speedway Special." It featured deluxe items such as larger tires, dual windshield wipers, sun visors, dual tail lamps, safety glass in all windows, and mohair upholstery, all at extra cost.

A Lasting Legacy

In 1940 the United States Army selected Willys Motors Incorporated to build their quarter-ton General Purpose (GP) four-wheel-drive vehicle because of the proven reliability of the Willys four-cylinder engine, and the extensive and well-equipped Willys manufacturing plants. During World War II the Willys "GP" vehicle was nicknamed the "Jeep." The Jeep was responsible for saving the lives of tens of thousands of American military personnel. It was so effective that even the Japanese (who had such a hatred and disdain for anything American) copied it for their own small military vehicles. The rest is history.

On September 30, 1943, Joseph W. Frazer resigned from Willys Motors Incorporated to become president of Graham-Paige Motors Corporation. Ward M. Canaday succeeded Frazer as president of Willys Motors. On June 8, 1944, Charles E. Sorensen became president of Willys Motors Incorporated, and Ward M. Canaday was elevated to chairman of the board. On July 1, 1946,

1943 Willys "G-P" (Jeep). (Source: American Motors Historical)

1943 Willys Jeep next to its "big brother." (Source: American Motors Historical)

1941 Willys-Overland "Americar." (Source: American Motors Historical)

1948 "Jeepster." (Source: American Motors Historical)

Sorensen and Canaday retired, and James D. Mooney became president and chairman of the board. Willys continued to build a civilian version of the Jeep, and introduced a new sporty "Jeepster."

In 1951 Willys Motors Incorporated introduced the "Aero-Willys," a new design by Clyde R. Paton attractively styled by Phillip O. Wright. Since the new Aero-Willys was merchandised along with the Jeep, the sales organization lacked sufficient dealer outlets: most dealers were dual with other makes so they did not push the sales of the Aero-Willys. During 1951 and 1952, Willys produced the engines for the Kaiser Henry-J. In 1954 Willys Motors Incorporated merged with the Henry Kaiser Corporation to form the Kaiser-Willys Corporation, which was eventually purchased by American Motors Corporation. After all other body

1948 "Jeepster." (Source: American Motors Historical)

*1952 Willys Aero Lark.
(Source: American Motors
Historical)*

*1954 Willys Aero Eagle
Deluxe. (Source: American
Motors Historical)*

types were discontinued, the Jeep vehicles were built and sold by the Jeep Corporation, a division of the American Motors Corporation. Then, on August 5, 1987, the American Motors Corporation merged with the Chrysler Corporation, becoming the Jeep-Eagle Division of the Chrysler Corporation.

Benjamin Briscoe

Why do some business enterprises succeed and others fail? Benjamin Briscoe was bright, energetic, and ambitious, and he did leave a mark on automobile manufacturing history, yet the pot of gold remained out of reach. Was it stubbornness, pride, poor business sense, or was it just that he managed to be in the wrong place at the wrong time?

* * *

Benjamin Briscoe was born in Detroit, Michigan, on May 24, 1867. His father, Joseph, and his grandfather, Benjamin, who settled in Detroit in 1837, were both prominent in the early railroad history of Michigan as inventors who improved railroad rolling stock. Therefore, it was only natural that the junior Benjamin, as well as his younger brother Frank, would be inclined toward making things that run on wheels.

The Early Ventures

In 1887, with only $432 in cash and an unlimited amount of faith and hope, Benjamin and Frank organized the Briscoe Manufacturing Company. They began as manufacturers of sheet metal items such as tubs, pails, cans, and just about anything else that could be made of sheet metal. They later branched out into the stove and steel range business. The company and plant grew to considerable proportions and was quite profitable.

During the storm and stress of the early years of automobile development one of the greatest difficulties was that of obtaining capital to start an automobile company. In 1902 the Briscoe brothers gave financial support to David Dunbar Buick to build his first car.

Also since the 1890s, when Haynes, Olds, and the Duryea brothers concerned themselves with automobile manufacture, it became apparent that one of the greatest challenges was how to construct a cooling coil or radiator.

In August 1903 Jonathan D. Maxwell developed a "thermo-syphon" cooling system and was able to convince the Briscoe brothers to manufacture the system (see the Maxwell chapter in Volume 4). The trio pooled their capital and, with some borrowed money, organized the Maxwell-Briscoe Motor Company on January 1, 1904. They leased the Tarrytown, New York, plant, buildings, and equipment, in which John Brisben Walker for a short while built the Mobile steam car.

Jonathan Dixon Maxwell

The experimental work on the first Maxwell car was begun in early 1904. Though difficult, Benjamin was able to generate phenomenal promotion and growth of the Maxwell-Briscoe Motor Company. The first year during the incorporation, 540 Maxwell cars were built and sold, 2,450 the second year, 3,540 the third year, 4,100 during 1907, 8,000 during 1908, and 16,000 during 1911. With the phenomenal sales, Benjamin Briscoe entertained many thoughts of expansion.

In the meantime, Alanson P. Brush left Cadillac Motor Car Company to join Frank Briscoe in late 1906 to organize the Brush Runabout Company. Frank not only had Benjamin's blessing but also his financial support. The first Brush runabout was introduced in January 1907. It had a vertical single-cylinder engine and a variable-speed clutch design. In May 1907 Brush designed a new 12-hp rated vertical two-cylinder engine having an ingenious crankshaft balancer, that was the forerunner of the "Harmonic Balancer" and the inertia balancer for radial aircraft engines.

Dreams of Expansion

Benjamin Briscoe, still eager for expansion, met with William C. Durant, who was enjoying great success with the Buick Motor Company, in Flint, Michigan on May 16, 1908. They discussed the possibility of forming an automotive combine of about 20 companies. After reviewing the pros and cons of the idea, Durant told Briscoe that he did not favor the proposal because he felt it was too big, as it involved too many concerns; but he would be willing to investigate a merger of a few successful automobile companies dedicated to volume production and sales. He recommended a merger of the Buick, Ford, Reo, and Maxwell-Briscoe companies.

Benjamin Briscoe, Henry Ford, William C. Durant, and Ransom E. Olds met approximately two weeks later in Detroit. The meeting was conducted in a well-organized manner and ended cordially, but without any

Benjamin and Frank Briscoe in a 1907 Brush runabout. (Source: NAHC)

decision. Briscoe was asked to arrange a meeting to include the four of them with the J.P. Morgan and Company banking firm; this was arranged a short time later in New York. At this meeting, Henry Ford started to have reservations about the proposed financial handling methods. He waited until the next scheduled meeting to shock Briscoe, Durant, Olds and the J.P. Morgan representative by stating that he would join the combine only on the basis of cash payment for his interests and would not accept stock as payment. Olds then said he, too, would demand cash payment for his interests. Because cash was not available, further discussion of the merger was fruitless, and the meeting was adjourned.

Durant did not want this opportunity to elude him and so he offered to deal with Benjamin Briscoe directly. The J.P Morgan and Company agreed to underwrite the floating of $5,000,000 of stock of a newly organized holding company called the International Motors. At the next meeting with the officers of the J.P. Morgan Company, Frederick L. Stetson, Morgan's counselor, began to question the plans by which Durant intended to acquire the capital, by exchanging Buick stock for stock in the new holding company. Mr. Stetson demanded a formal meeting of all Buick stockholders. Durant, having full control of Buick, insisted that it would not be necessary because he owned all the Buick stock or had proxies representing 100%. But Mr. Stetson stormed out, ending the meeting. As a result J.P. Morgan and Company backed out, losing the opportunity to pioneer the financing of a growing automobile industry.

Durant was still so engrossed with consolidation that he decided to go it alone. On September 16, 1908, he organized and incorporated the General Motors Company.

The United States Motor Company

Created for the Wrong Reasons?

Benjamin Briscoe, not wanting to be outdone by Durant, decided to arrange his own automobile combine, the United States Motor Company. Unfortunately, successful automobile manufacturers were not beating down his door. Durant and General Motors skimmed off the profitable companies and the remaining successful companies were simply not interested.

Benjamin should have been satisfied with the profitable Maxwell-Briscoe Motor Company, but his compulsion drove him to merge with the Columbia Motor Car Company in February 1910. Perhaps the reason Columbia looked attractive to Briscoe was that it was owned by a group of New England financiers, the same group that backed him when he founded the United States Motor Company.

The Columbia Motor Car Company was a spin-off of Pope Manufacturing and was incorporated in the state of Connecticut on June 30, 1909. During its organization it acquired the assets and properties of the Electric Vehicle Company, which included the rights to the Selden Patent that seemed lucrative at the time. Little could Briscoe anticipate the outcome of the patent infringement suits that would be lost in court during 1911, causing the Selden Patent to become worthless.

On April 14, 1910, Briscoe bought the assets of the Alden Sampson Manufacturing Company of Pittsfield, Massachusetts, and moved the operations to a suburb of Detroit (Highland Park) Because he was in need of a lower-priced car in the United States Motor Company model lineup, Benjamin brought in the Brush Runabout Company, still run by brother Frank. (The Sampson and Brush plants were just two blocks apart and later became important to Maxwell Motor Corporation and the Chrysler Corporation operations in Michigan.)

1910 Maxwell AA runabout.
(Source: NAHC)

After acquiring Brush, the United States Motor Company purchased the adjoining properties and arranged for the grouping of its Detroit plants. These plants included the Brush Runabout Company, Briscoe Manufacturing Company, Gray Motor Company, and the light car department of the Alden Sampson Manufacturing Company. By this time, Maxwell-Briscoe Motor Company was already in operation in a new plant in Newcastle, Indiana. (The New Castle plant is still important in producing forgings for the Chrysler Corporation.)

Finally, desperate to fill out its lineup, United States Motor Company acquired the Dayton Motor Car Company on June 14, 1910, makers of the Stoddard-Dayton and Courier automobiles. The Stoddard-Dayton was a fine automobile but very expensive, and the Stoddard-Dayton-Knight would be of no help either in attracting the everyday buyer.

The Inevitable Failure

With numerous plants in five different states, the United States Motor Company was having difficulty controlling costs, and communication was almost impossible. While Maxwell-Briscoe, the Brush Runabout Company, and Briscoe Manufacturing Company were operating profitably, the others were not and accumulated great losses.

The demise of the American Bicycle Company and the impending bankruptcy of the Pope Manufacturing Company combine should have been a warning signal to Benjamin Briscoe. However, history has a habit of repeating itself, and he was intent on trying to outdo Durant.

All during 1911 glowing reports were made regarding the success of United States Motor Company. This was firmly emphasized during the first annual gathering of company officials and sales representatives during mid-summer 1911 at Cedar Point, Ohio. They boasted that United States Motor Company had 14,000 employees and would have an annual output of 53,000 cars, which ranged in price from $485 to $5,000. The exaggerated promise was partially justified as Maxwell-Briscoe sales were approximately 18,000 cars and Brush runabout would be over 20,000 cars. However, the Brush runabout was a low-profit car.

1910 Stoddard-Dayton 10-F limousine. (Source: NAHC)

So confident were the United States Motor Company officials that on August 24, 1911, they boldly introduced a more expensive Stoddard-Dayton-Knight with a six-cylinder Knight engine. It was in mutual company with the already expensive Columbia-Knight priced at $4,900 (later $6,000). The prospective car buyers stayed away from Columbia, Stoddard-Dayton, and Sampson in droves. In spite of the optimistic claims of success, the United States Motor Company was definitely in a serious sales slump and deep financial trouble.

The organization of the United States Motor Company started to unravel in early 1912, when Charles G. Stoddard resigned on February 1, 1912. When he left he was able to retain the license rights to the Knight engine, which he passed on to H.J. Edwards, who left United States Motor Company with Stoddard.

After that the company went into total disarray. Attempts to bring the E.R. Thomas Motor Company into the United States Motor Company combine failed when E.R. Thomas went into receivership. Frank Briscoe resigned on August 8, 1912, and by September 11, the United States Motor Company was in the hands of its receivers. Benjamin Briscoe resigned on October 2, effective November 15, and Jonathon D. Maxwell resigned on December 22.

Judge Charles M. Hough was authorized to receive bids for the properties and assets of the United States Motor Company and, on January 14, 1913, accepted the bid of $7,000,000 by the Standard Motor Company, which was incorporated under Delaware laws on January 11, 1913. The officers of Standard Motor Company were: Walter E. Flanders president, W.F. McGuire vice president, Carl Tucker treasurer, and M.L. Anthony secretary. However, the name of Standard Motor Company was conflicting; thus, the name was changed to Maxwell Motor Company Inc. on February 6, 1913.

Starting Again

In early 1913, Benjamin and Frank Briscoe traveled to France. While there, they formed the Briscoe Freres at Bellancourt to build the Freres cycle-car and light car designed by French engineers. Back in the States, in March 1914, they organized the Argo Motor Company to merchandise the Freres cycle-car, manufactured in France.

1911 Stoddard-Dayton toy tonneau. (Source: NAHC)

Also, in late 1913, the Briscoe brothers formed a new Briscoe Motor Company in Jackson, Michigan, in order to build a car bearing their name. Because their credit rating was not AAA on Wall Street, they obtained financial backing in Chicago.

The Briscoe chassis was built in the plant of the Lewis Spring and Axle Company in Jackson, Michigan, while assembly took place in the plant machine shop they purchased from the Jackson Machine Parts Company. The 1914 Briscoe car was introduced at the National Auto Show in New York in January 1914. It had a 107-in. wheelbase chassis, powered by a four-cylinder, 33-hp, L-head engine. Available as a touring and roadster model, it was priced at $785.

Although the Briscoe car was similar to other makes of the same size, it had one identifying feature: the "Cyclops" single headlamp mounted in front at the top of the radiator shell. While this feature certainly made the car unique, it also made it difficult to sell. The Cyclops headlamp stuck around through 1915, but then gave way to two conventional headlamps.

The 1916 Briscoe models were introduced on July 22, 1915, and offered a choice of either four-cylinder, 33-hp, L-head engine or the new Ferro valve-in-head V-8 engine available in the same 114-in. wheelbase chassis, in a touring car or "cloverleaf" roadster model priced at $950 with the V-8 engine, and $750 with the four-cylinder engine. Briscoe further offered the buyer of the four-cylinder model the option of having the engine replaced with a V-8 within 30 days at a price of only the difference of $200 and modest labor costs. There weren't too many takers. The Briscoe car was not favorably received, and production problems mounted.

Shortly after war was declared against Germany in 1917, Benjamin Briscoe felt that the Briscoe Motor Company was operating satisfactorily and that it was being managed by competent subordinates, so he joined the United States Navy. He had the responsibility for directing the maintenance of vehicles used by the military staff, and the servicing of naval aircraft. He was promoted to Commander in 1918 and was awarded the Navy Cross upon his release to inactive duty.

Calling it Quits

After the Armistice on November 11, 1918, Benjamin returned to discover that the Briscoe Motor Company has not been so well managed after all. With all the stress and financial burden, Briscoe for once realized "when enough was enough." In early 1921, he appointed Clarence A. Earl (vice president of Willys Overland until 1920) as president of Briscoe Motor Company.

Because Benjamin Briscoe was anxious to get out of the automobile manufacturing business, and Clarence A. Earl was egotistical enough to pay the price to build a car bearing his name, Briscoe unloaded the company on Earl in October 1921. During the existence of Briscoe Motor Company, approximately 32,000 Briscoe cars had been built.

The Briscoe Motor Company became Earl Motors Incorporated, and the Briscoe car became the Earl during 1922 and 1923 when approximately 2,000 Earl cars were built. Benjamin also sold the Briscoe sheet metal manufacturing business to Continental Can Company.

After 1921, Benjamin Briscoe certainly was not idle. He was engaged as a "wildcatter" in the oil well business, in gold mining in Colorado and Utah, and in iron ore mining, smelting, and steel mills. He finally retired in 1938 to his large plantation in Florida. Benjamin Briscoe passed away on June 26, 1945, after a long, energetic and exciting life of 77 years.

Charles Matheson

Charles Matheson's life was anything but dull, and it cannot be said that he didn't try. But, even with the turbulence, frustrations, and disappointments, he came through it all. He just happened to be at the wrong place at the wrong time. Nevertheless, he certainly had a great impact on the automotive industry, and left a legacy of the bridges he built for others to follow.

* * *

Charles W. Matheson was born in Grand Rapids, Michigan, in 1877, and graduated from Central High School in 1895.

During 1893 Matheson was employed by the Frederick Macey Company of Grand Rapids, Michigan, where he remained for nine years. During the last three years of his association with Macey he became a partner in charge of the purchasing, advertising, and sales for the company. Then in 1902 Matheson severed his connection with the Fred Macey Company, and with his brother, Frank F. Matheson, organized the Matheson Motor Car Company in 1903.

He was disappointed with the operations in Grand Rapids, as only six cars were built during 1903, so in early 1904 he searched for a new location. He chose Holyoke, Massachusetts, not so much for its location and a favorable plant, but for the fact that the services of Charles R. Greuter, a brilliant engine design engineer, came with the deal. During the balance of 1904 the Matheson engines were built in Holyoke, and the cars were assembled in Grand Rapids. Finally in 1905 the car assembly operations were moved to Holyoke, employing about 150 people.

In late 1905 Matheson was offered a deal for a factory by the Wilkes-Barre, Pennsylvania, Board of Commerce which he felt he could not refuse, so operations were moved there in early 1906. The move involved twenty carloads of machinery, equipment, and subassemblies. About fifty employees and their families moved to Wilkes-Barre with him.

The new 1906 Matheson car featured a unique, Greuter-designed, four-cylinder engine with overhead valves actuated by rocker levers and a camshaft located on the head alongside the valves. The Matheson engine proved its performance by establishing a new record at the Atlantic City Speedway in September 1906. Carrying a full load of seven adult passengers, it covered the measured mile in 50 seconds, a speed of 72 mph. Unfortunately, the price tag of the car was too high and the lack of sales confirmed it.

Recognizing the need for a lower-priced car, Matheson added a smaller car powered by a Greuter-designed, four-cylinder overhead camshaft engine. Even with the downsizing, the price was reduced by only $1500.

	1906	1906
Wheelbase (in.)	118	112
Price	$7,500	$6000
No. of Cylinders / Engine	4	4
Bore x Stroke (in.)	5.50 x 6.00	4.50 x 6.00
Horsepower	60/65 adv, 48.4 ALAM	40/45 adv, 32.4 ALAM
Body Styles	touring	touring
Other Features		

The two Matheson car lines for 1907 consisted of the Four-35 and the Four-50. The Wall Street Panic of October 22, 1907, didn't help sales as all businesses slowed to a crawl.

	1907 Four-35	1907 Four-50
Wheelbase (in.)	123	128
Price	$4,500-5,500	$5,500-6,500
No. of Cylinders / Engine	4	4
Bore x Stroke (in.)	4.50 x 6.00	5.50 x 6.00
Horsepower	40/45 adv, 32.4 ALAM	50 adv, 48.4 ALAM
Body Styles	seven-passenger touring, limousine landaulet, runabout	
Other Features		

Although the cars were built in Wilkes-Barre, the wholesale merchandising was through the Matheson Automobile Company at 1886-88 Broadway in New York City. Retail sales for the Matheson car were handled by the Palmer and Singer Manufacturing Company who were also agents for the Simplex and Isotta-Fraschini automobiles. On August 28, 1907, Palmer and Singer took over the wholesale merchandising of the Matheson cars; Matheson of Wilkes-Barre built the chassis for Palmer and Singer cars.

On May 30, 1907, a stripped-down Matheson won the Wilkes-Barre "Devil's Despair" Hill Climb in 1 minute and 59.4 seconds.

The 1907 Matheson Four-50 was carried over into 1908 with the prices and technical specifications remaining the same. The Matheson Four-35 was discontinued. During 1908 Charles Greuter resigned from the Matheson Motor Car Company and was succeeded by L.D. Kenan.

The Matheson Four-50 was continued into 1909, but the prices were reduced by $1,000. A new Matheson Six-50 was added, featuring Greuter's masterpiece, a six-cylinder overhead camshaft engine. Despite the relatively low price of the Six-50, the 1909 sales did not improve.

	1909 Four-50	**1909 Six-50**
Wheelbase (in.)	128	125.5
Price	$4,500-5,500	$3,000
No. of Cylinders / Engine	4	6
Bore x Stroke (in.)	5.50 x 6.00	4.50 x 5.00
Horsepower	50 adv, 48.4 ALAM	50 adv, 48.6 ALAM
Body Styles	roadster, seven-passenger touring, limousine	five-passenger touring
Other Features		

The 1910 Matheson Four-50 was designated as "50," accompanied by price increases to $5,000 to $5,750. The 1910 Matheson Six-50 became known as the "Silent Six." Offered in seven body styles, the Silent Six was priced at $3,500 for the touring to $4,700 for the seven-passenger limousine.

During 1910 low car production and sales were not the only problems for Matheson Motor Car Company. On July 7th, Matheson applied to the courts of Luzerne county to appoint a receiver. The three receivers appointed were Col. Asher Miner, president, W.C. Shepherd, Matheson director, and Harold L. Pope, Matheson designer/engineer. On July 11, 1910, the receivers obtained permission from the court to borrow $50,000 to pay the employees and to purchase the materials necessary to complete the cars in process.

On November 17, 1910, the Matheson Automobile Company of New York took over the Matheson Motor Car Company of Wilkes-Barre, Pennsylvania. This action was authorized by the Common Pleas Court of Luzerne County, Pennsylvania, and terminated the receivership of Matheson Motor Car Company. W.C. Shepherd was elected president, J.W. Hollenback vice president, E.F. Matheson secretary, and Henry H. Pease treasurer. The 1911 Matheson cars were a carryover of the 1910 models with the prices and specifications remaining the same.

For 1912 Matheson proliferated into twelve body styles in the Silent Six, including four new body types on a 135-in. wheelbase chassis. The 50 was discontinued.

1910 Matheson Model E special speedster.

*1910 Matheson Model E
seven-passenger touring.*

*1910 Matheson Model M
toy tonneau.*

	1912 Silent Six	**1912 Silent Six**
Wheelbase (in.)	125.5	135
Price	$3,500-4,700	$4,500-6,700
No. of Cylinders / Engine	6	6
Bore x Stroke (in.)	4.50 x 5.00	4.50 x 5.00
Horsepower	50 adv, 48.6 NACC	50 adv, 48.6 NACC
Body Styles	8	4
Other Features		

Because financial conditions did not improve, the business, title, equipment, and property rights of Matheson Automobile Company were ordered to be sold by the receiver. On the first day of the sale on May 20, 1913, the total money realized from the sale was $81,000 for the goodwill, trademarks, patterns, drawings, etc. The sale continued until the rest of the assets were sold. Frank F. Matheson remained in Wilkes-Barre during the sale.

In the meantime, Charles Matheson became western manager for the Palmer and Singer Manufacturing Company, having charge of all the territory west of Pittsburgh, Pennsylvania. Later he became general sales and advertising manager for Palmer and Singer, a position he held until the fall of 1914.

In November 1914 Charles Matheson joined the Dodge Brothers, and was assigned the eastern half of Pennsylvania and New York and all of New Jersey and Connecticut. When A.I. Philip resigned in 1921 to join the Durant Motors Incorporated, Matheson moved up to general sales manager, and then vice president of sales.

In March 1924 Matheson was added to the General Motors Corporation Executive Staff as assistant to president Alfred P. Sloan. In September of that year he became vice president and director of sales for the Oakland Motor Car Company. Since 1923 Oakland and Oldsmobile experienced a lack of sales, not due to anyone's personal abilities, but rather because of poor product value and lack of reliability.

Feeling that he was being blamed for Oakland's dismal sales performance, and operating in an unfriendly environment, he resigned as vice president of sales of the Oakland Motor Car Company on December 28, 1926. The overwhelming sales success of the new 1926 Pontiac confirmed that the lack of Oakland sales was the product itself.

Matheson was not idle very long though, because immediately upon his resignation from Oakland, he was offered the vice presidency of the Kelvinator Corporation, which he accepted. He stayed with the Kelvinator Corporation until he was prevailed upon by K.T. Keller and Byron C. Foy in March 1928 to join the Chrysler Corporation in a newly created executive position. By mid-1928, Matheson was named vice president of sales for the DeSoto Motor Corporation. Byron C. Foy was elected vice president of Chrysler Corporation and president of DeSoto. The new DeSoto made its debut on August 4, 1928, and set a sales record for a new car during its first twelve months, selling 81,065 units.

Matheson resigned from the DeSoto Motor Corporation on October 28, 1930, to become general sales manager for the Graham-Paige Motors Corporation. At that time the general sales manager reported directly to the executive vice president, F.R. Valpey. A.I. Philip also joined Graham-Paige as sales counselor. Having great expectations that the new Graham "Blue Streak" would be a sales success in 1932, all effort was put forth to make the buying public aware of the many advancements and innovations the Blue Streak had to offer.

But, all the product offering and sales effort became futile, since 1932 turned out to be the lowest level of the cruel Depression, and many well-established automobile manufacturing companies went bankrupt or went into receivership. Graham-Paige trudged along, even into farm tractor manufacture, with modest success. Matheson was on the Graham Brothers' special assignments during the turbulent 30s. (See the full story of Graham-Paige in the chapter on the Graham brothers in Volume 2.)

Charles W. Matheson died on August 12, 1940, as the result of an automobile accident near Brodhead, Wisconsin, while he was on Graham company business.

Allison/Fisher/ Newby/Wheeler and the Indianapolis Motor Speedway

Fisher, Wheeler, Newby, Allison.
(Source: Indianapolis Motor Speedway Corp.)

Why was Indianapolis destined to be the automobile racing capital of the world, providing a great sporting spectacle on the greatest race course in the world? What would prompt three successful businessmen and a great bicycle and balloon racing entrepreneur to meet at a restaurant in 1909 and decide to risk their fortunes and reputation on an automobile race track? James Allison and Carl Fisher had formed the successful Prest-O-Lite Company in 1904, Arthur Newby was the president of the successful National Motor Vehicle Company, and Frank Wheeler was president of the Wheeler-Schebler Carburetor Co. Was it an act of "Divine Providence" that brought these daring men together to form the Indianapolis Motor Speedway Corporation and build the now-legendary 2-1/2-mile Speedway?

* * *

The Founders

James A. Allison and Carl G. Fisher

James A. Allison, son of Noah S. and Myra J. Allison, was born on August 11, 1872, in Marcellus, Michigan. While James was still a toddler, he and his father, mother, four brothers, and one sister moved to South Bend, Indiana, and then, in 1880, to Indianapolis. His father was a printer by trade and established the Allison Coupon Company, later inherited by the sons in 1890.

Carl Graham Fisher was born in Greensburg, Indiana, in 1874. Leaving school at an early age, he went to Indianapolis and became a bicycle salesman, through which he soon discovered his strong sales and

James A. Allison. (Source: Indianapolis Motor Speedway Corp.)

promotion skills. After two or three years he established his own bicycle agency, but, like many others engaged in the bicycle industry, he foresaw the collapse of the bicycle industry with the coming of the motorcycle and automobile. He purchased an Orient motorcycle and gave exhibitions at county fairs and other forms of public crowd excitement. As the popularity of steam cars grew, he obtained a distributorship for the Mobile steam car, of which only a few were built and sold. He then established agencies for gasoline-engine-powered cars, including his favored Stoddard-Dayton. Many other dealers followed his lead, and thus the famous "Automobile Row" on Capitol Avenue was established.

In the early days of the automobile industry, the headlamps used were of the acetylene type, following the practice of the bicycle industry. The acetylene used by the lamps had to be produced by an acetylene generator using calcium and carbide, carried on the car. This system was unsatisfactory because of the inability to regulate the gas production on rough roads, and the messiness of the gas generator. Early in the 20th century, it was discovered in France that large quantities of acetylene gas could be dissolved in acetone and compressed for storage or use. This gave rise to the dissolved-acetylene industry. James Allison and Carl Fisher developed the process in Indianapolis in 1904, and organized, with P.C. Avery, the Concentrated Acetylene Company to manufacture and fill acetylene cylinders. P.C. Avery left the company in 1906, at which time Allison and Fisher renamed the company the Prest-O-Lite Company. From a humble beginning, it is almost impossible to comprehend the phenomenal growth of the Prest-O-Lite Company. By 1913, when automobile lighting became a function of the electrical system, Prest-O-Lite had become a common household word, as many new uses for acetylene gases were discovered. They had erected a new plant in the township of Speedway Indiana to keep up with the demand. Prest-O-Lite Company became the largest producer of acetylene and other products, with a nationwide system of charging and a worldwide distribution of its products.

By 1917, recognizing the potential for further growth, the Prest-O-Lite Company joined with Electro Metallurgical Company, the Linde Air Products Company, the National Carbon Company Incorporated, and the Union Carbide Company, to form the Union Carbide and Carbon Corporation.

Meanwhile, as will be described in more detail later in this chapter, four prominent Indianapolis men including Fisher and Allison (the others being A.C. Newby and Frank H. Wheeler), all of whom fortune had smiled upon through the medium of the automobile industry, joined together in 1909 and decided to build a motor speedway near Indianapolis. The Indianapolis Motor Speedway was a 2-1/2-mile race course, rectangular with curved corners, built at West 16th Street and Georgetown Road. Carl Fisher was president, James Allison vice president, Arthur Newby secretary, and Frank H. Wheeler treasurer.

In order to participate in racing events, Allison organized the Indianapolis Speedway Team Company on September 14, 1915. The engineer-mechanic employees were engaged in the modifying and rebuilding of the racing cars. During 1917 the Indianapolis Speedway Team Company received worthy contracts from the

United States Army Air Corps for the reconditioning and rebuilding of Liberty aircraft engines. The company had rebuilt approximately 3,000 Liberty aircraft engines by November 11, 1918, and was renamed Allison Engineering Company.

On June 12, 1923, Fisher resigned from the Indianapolis Speedway Company to pursue his many interests, especially in the development of Miami Beach, Florida. Allison bought Fisher's interest, and was elected president of the Indianapolis Speedway Corporation, Arthur Newby was elected vice president, and Theodore E. Myers was named secretary-treasurer.

On August 17, 1927, Allison sold his controlling interest in the Indianapolis Motor Speedway to Edward V. Rickenbacker, a former racing driver of great renown (see the Rickenbacker chapter in this volume). The following year Rickenbacker arranged for the sale of the Allison Engineering Company to the Fisher Brothers Foundation, who in turn sold it to the General Motors Corporation, where it became the Allison Engineering Division, and continued to develop aircraft engines and geared drives.

Carl G. Fisher. (Source: Indianapolis Motor Speedway Corp.)

During the 1930s Allison Engineering Division developed the famous V-1710 Allison V-12 aircraft engines, which powered the famous Lockheed P-38, the Curtiss P-40 and many other combat aircraft of World War II. With the return of peacetime in 1946, Allison aircraft engines continued to make history in the 'unlimited' powerboat racing craft, such as Miss Pepsi and many others.

Unfortunately James Allison did not live long enough to witness the fruition of all his aircraft engine endeavors. While on a trip to New York, he developed a severe cold. It got progressively worse, and while returning to Indianapolis, it turned into pneumonia. When he arrived in Indianapolis, he had already gone into a coma, and he died on August 3, 1928. While his life was cut prematurely short at age 55, he accomplished many of his goals, for he left a legacy for all Americans to benefit from.

Carl Fisher's promotional ideas, backed by his initiative and drive, made it possible for him to become a millionaire before the age of 21. Before retiring as president of the Indianapolis Motor Speedway, he had already embarked on a gigantic enterprise in Florida. It was during the winter of 1913-14, that Fisher, from the deck of his yacht anchored at the Miami docks, envisioned the opportunity of developing a beach resort on the large "key" across Biscayne Bay from Miami, Florida. In February 1915 the *Indianapolis News* declared that "Carl G. Fisher, head of the Indianapolis Motor Speedway, is establishing a great resort for winter sports, across the Biscayne Bay opposite Miami, Florida."

It went on to say that Fisher had invested a large amount of money into a huge sand pile on a key he had purchased, without any financial help from anyone. Fisher has developed on his island a magnificent 18-hole golf course, surrounded by thriving trees and grass. They patiently planted approximately 50,000 coconut and Australian pine, and tended to their care. The reclaiming of the marshy swamp was accomplished by pumping sand and marl from the bay, for over a year and a half, enlarging the key by approximately 300 acres, adding to the 400 acres of the key itself. By removing the sand and marl from the channel, they created a harbor 2-1/2 miles long and up to 1/2-mile wide, which could accommodate any yacht in the United States. Fisher had a causeway built to connect Miami City to the long key which would be called Miami Beach.

Carl Fisher literally created Miami Beach, upon which a historical marker quotes: "He carved a Great City out of a Jungle." Along with Henry M. Flagler, they developed the systems and accommodations for travel in Florida and the Keys. Fisher witnessed the reconstruction of Flagler's overseas highway, between the Keys in 1938, after a hurricane destroyed the railroad tracks in 1935. Carl Fisher passed away in his home in Miami Beach, Florida, on July 15, 1939, at age 65.

Arthur C. Newby

Arthur C. Newby. (Source: Indianapolis Motor Speedway Corp.)

Arthur C. Newby was born on December 29, 1865, in Monrovia, Indiana, a small village southwest of Indianapolis. Shortly after his birth, his parents and family moved to Kansas City, Missouri, and later moved to California. When Arthur was about ten years old the family came back to Indianapolis. Arthur's first job was at the Dickson Trade Palace where he earned $1.50 per week. By 1882 he became office boy, timekeeper, assistant bookkeeper, and finally a bookkeeper at the Nordyke and Marmon Company, manufacturers of grain milling equipment.

In December 1890, he joined with E.C. Howe and Edward O. Fletcher to establish the Indianapolis Chain and Stamping Company. This happened at the peak of the bicycle's popularity and the rapidly growing bicycle industry. In the manufacture of bicycle components one of the items in great demand was bicycle drive chains, and at one point Newby's company supplied over 50% of the bicycle chains manufactured in America. The very apparent success of the company spurred the envy of competitors and bicycle manufacturers, and provided Newby with "an offer he couldn't refuse." He sold his firm to the American Bicycle Company which, later after reorganization, became the Diamond Chain Company, one of the foremost timing chain suppliers to the automobile industry.

After he established the "Newby Oval," a 1/4-mile board race track for bicycle racing, Newby realized that the motorcycle and automobile would replace the bicycle. He joined with Albert E. Metzger, Charles E. Test, L.S. Dow, and Philip Goetz, formerly of the American Bicycle Company, to form the National Automobile and Electric Company to build electric-powered vehicles. It was incorporated on February 28, 1900, with a capitalization of $250,000. While National started out for electric-powered vehicles, by 1906 it began manufacturing gasoline-engine-powered cars. Newby was one of the directors at first and then was elected president of the renamed National Vehicle Company in 1906. The National Company built cars powered with two-, four-, six-, and by 1912, twelve-cylinder-engine cars. The National "Highway" Twelve was considered one of the finest cars of its day.

It was only natural that when the Indianapolis Speedway Company opened the track in 1909, National would be in the race. National entered three racing cars, driven by Charles Merz, John Aitken, and Thomas Kincaid. After the track was paved in 1910, National entered three racing cars for the 1911 Indianapolis "500" race, with Charles Merz, John Aitken, and Howdy Wilcox as the drivers. Charles Merz finished in seventh place. In 1912 National again had three entries, driven by Joe Dawson, Howdy Wilcox and David Bruce Brown. Joe Dawson went on to win the race at 78.72 mph.

In 1909 James Allison, Carl Fisher, Arthur Newby, and Robert Hassler, chief engineer of National Motor Vehicle Company, organized and incorporated the Empire Motor Car Company, to build the Empire car, an agile, very attractive "runabout" resembling Mercer, Oldsmobile, Hudson, Cole, and many other popular models. It sported racing-type bucket seats, a round fuel tank placed transversely behind the seats, a 96-in. wheelbase chassis, 20-hp four-cylinder engine, and was priced at only $800. The Empire could have been successful if the founders' interest had not been stolen by the Indianapolis Motor Speedway, which in the long run proved to be a more successful project. The Empire interests were sold to another group of Indianapolis businessmen in 1911, and by 1919 was discontinued.

In 1916, Newby retired as president of National Motor Vehicle Company, and was succeeded by G.M. Dickinson. After his retirement, Newby spent much of his time in Florida, where he was well-renowned for his yachting and powerboat activities. But, most of all, Newby was well known and loved for his "selfless" philanthropy. He shared most of his earned wealth with numerous health charities and universities. He did much for the conservation of wildlife in the state of Indiana. The only stipulation of his generous contributions to private charities was that the giver remain anonymous. He died on September 11, 1933, leaving a void hard to fill.

Frank H. Wheeler

Frank H. Wheeler was born in Manchester, Iowa, on October 24, 1864, but moved to California during his childhood. After his formal education, Frank was engaged in several occupations and enterprises. It had

Frank Wheeler. (Source: Indianapolis Motor Speedway Corp.)

been reported that he had made and lost two fortunes before coming to Indianapolis as a traveling salesman. It was his good fortune that during his business contacts he met George Schebler, an inventor of a successful carburetor for gasoline engines, and that during their subsequent meetings their ideas, methods, and habits were compatible.

In 1907 they organized and incorporated the Wheeler-Schebler Carburetor Company. They had to struggle for a short while, but the abnormal prosperity and growth of the automobile industry created a great demand for their quality carburetor, which by this time had earned a reputation of high esteem by the automobile industry. The Schebler carburetor became one of the best-known products in the world. (The Wheeler-Schebler name was awkward to pronounce, hence the name of Wheeler was omitted.) George Schebler, pleased with their success, in 1915 decided to retire. He sold his interest in the Wheeler-Schebler Carburetor Company to Frank Wheeler for a reported $1,000,000.

Wheeler continued as president of the Wheeler-Schebler Carburetor Company through 1921, but he sold his shares in Indianapolis Motor Speedway to James Allison in 1917. He had suffered from diabetes during the latter part of his life, and in early 1921 sustained a foot injury which developed into gangrene. Frank Wheeler died on May 27, 1921.

After the passing of Wheeler, the company continued to build Schebler carburetors well into the 1930s, when it merged with the Marvel Carburetor Division of General Motors Corporation. The Marvel carburetor later became the Rochester carburetor, which after the coming of electronic fuel injection, was absorbed into the AC-Delco division. After his passing, Frank Wheeler is remembered not only for his industrial accomplishments, but mostly for his generous philanthropy. There was no limit of endowments to numerous children's charities and children's recreational facilities.

The Indianapolis Motor Speedway

On February 8, 1909, the articles of incorporation were filed for the Indianapolis Motor Speedway Company, capitalized at $250,000. Carl G. Fisher was elected president, Arthur C. Newby first vice president, Frank H. Wheeler second vice president, and James A. Allison secretary-treasurer. Fisher and Allison held the controlling stock. Through Lemon H. Trotter, a real estate speculator, they agreed upon was the Pressley farm, then owned by the Munter and Chenoweth families, consisting of 1/2 section (320 acres) of land purchased at $200 per acre, and later 80 acres of adjoining land was purchased at $300 per acre.

Carl Fisher was given the task of designing the shape and dimensions of the 2-1/2-mile automobile racing track. At a meeting at his home, Fisher was able to show Allison, Newby, and Wheeler a rough drawing of his speedway plan, and in his back yard he made a miniature model of the track, using lime emulsion (whitewash) to outline the proposed race course. The preliminary drawings included a road course inside the

Carl Fisher and the Speedway mockup. (Source: Indianapolis Motor Speedway Corp.)

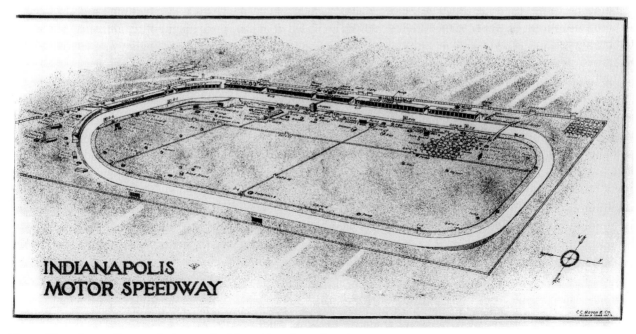

INDIANAPOLIS
MOTOR SPEEDWAY

Plan of Speedway. (Source: Indianapolis Motor Speedway Corp.)

Carl Fisher, first president of the Indianapolis Motor Speedway, is shown at the wheel of the car on the far left during the first inspection tour of the grounds by representatives of the press on May 1, 1909. James A. Stuart, who later succeeded B.F. Lawrence as managing editor of the Indianapolis Star, is in the rear seat of the far right car. Lawrence is sitting in the passenger seat of that car. Paul Willis, first auto editor of the Star, is at the wheel of the middle car. (Source: Indianapolis Motor Speedway Corp.)

2-1/2-mile rectangular track with quarter-circle turns at each corner. The physical dimensions of the rectangular track are: the complete circuit (3 ft. from the inside edge) is 2-1/2 miles in length; the front straightaway and back stretch are 3,301 ft. long; the short chutes (end straights) are 660 ft. long, and the quarter-circles at each corner are 1,320 ft. along the course line 3 ft. from the inner edge; the turns have a radius of 840 ft., and the first 50 ft. of the width of the turn from the inner edge is banked at 16 degrees and 4 seconds, and the remaining 10 ft. of width is banked at 36 degree and 40 minutes. The front straightaway and back stretch are 50 ft. wide, the turns not including the safety aprons, are 60 ft. wide.

Fisher opened the speedway, even before it was finished, by staging the first national balloon race held in the United States. Nine balloons (50 ft. in diameter) participated. Carl Fisher personally, with George Bumbaugh, a famous balloonist, rode in the bright yellow "Indiana" balloon, which finally landed in Virginia. It was reported that 75,000 spectators witnessed the event.

After months of toil and worry about the plot which was once a field of golden grain, the dreams and hopes of the founders of the Indianapolis Motor Speedway were finally about to be realized with the inaugural races to begin on Thursday, August 19, 1909. There were 60 cars entered, in five categories of 1 mile, 5 miles, 10 miles, 250 miles, and a 300-mile event scheduled for Saturday, August 21, 1909. Barney Oldfield won the 1-mile event against time, at 43.1 seconds. Louis Schwitzer won the 5-mile event the first day in a Stoddard-Dayton. The 5-mile race for stock chassis with 301- to 450-cu.-in. displacement was won by Wilfred A. Bourque in a Knox car. The 10-mile free-for-all was won by Harry Stillman in a Marmom. Bob Burman won the 250-mile race in a Buick racing car. The 300-mile race scheduled for Saturday, August 21, was stopped at the 235-mile mark due to the number and severity of accidents (several fatal).

The balloon race to open the Speedway. (Source: Indianapolis Motor Speedway Corp.)

Up until that time the track surface was not paved, and the dirt surface was treated with oil to keep the dust down. The fatal accidents during the first races so impressed the management of the inadvisability of racing on a dirt track, that they decided to halt all racing until they could resurface the 2-1/2-mile track. There were 3,200,000 paving bricks used, grouted with cement. At the center of the starting line there was a brick molded of bronze. Before the resumption of racing a rumor was circulated around Indianapolis that someone stole the "Golden Brick." It was subsequently proven that it was a hoax, probably a publicity stunt.

The Speedway resurfaced with bricks. (Source: Indianapolis Motor Speedway Corp.)

After the Indianapolis Motor Speedway reopened in 1910, five racing events were held, three in May and two in July. Tom Kincaid won the 100-mile race on May 27, 1910, in a National race car. Ray Harroun won the 200-mile race on May 28, 1910, in a Marmon, and the 50-mile race on May 30, 1910, also in a Marmon. Racing was again resumed on July 2, 1910, with Bob Burman winning the 100-mile event in a Marquette-Buick. The 200-mile race on July 4, 1910, was won by Joe Dawson in a Marmon. The racing cars of that period were in fact stock production models, with the fenders, running boards, lights, and windshield removed.

After the racing events in 1910, the Indianapolis Speedway management decided on a single 500-mile race to be run on Memorial Day, May 30, 1911. The 500-mile distance was chosen because it allowed a full day of racing, and yet it was not too long, and the date of May 30th was selected because it was midway between corn planting and harvesting seasons. The starting positions were determined by the sequence of entries. The front row of the starting cars were: (1) Lewis Strang in a Case, (2) Ralph DePalma in a Simplex, (3) Harry Endicott in an Interstate, and (4) John Aitken in a National. Forty of the 46 entries were in the starting field, as no qualification timing runs were made. Carl Fisher drove the Stoddard-Dayton pace car to start the race.

On the morning of May 30, 1911, The Indianapolis 500 race started on time, with the detonation of the starting bomb. The first lap was a pace lap, and did not count, but on the next lap the green flag was dropped, and the racing cars were on their way. The first mishap occurred on lap 12, when Arthur Greiner's car hit the southeast wall (at turn 3). Greiner suffered a broken shoulder, but his mechanic Sam Dickson was killed. After that, positions changed frequently, and on the straightaway during lap 87, Joe Jagerberger's mechanic fell from the speeding race car. Harry Knight in a Westcott swerved to avoid hitting the mechanic, and crashed into Herb Lytle's Apperson in the pit area, then flipped into Caleb Bragg's Fiat. Knight suffered a fractured skull, and Lytle's Apperson was totally demolished, which prompted Apperson to withdraw from racing.

The start of the 1911 Indianapolis 500 race. (Source: Indianapolis Motor Speedway Corp.)

Ray Harroun, driving a Marmon (#32), won the 1911 Indianapolis 500. (Source: Indianapolis Motor Speedway Corp.)

The car's positions changed as the race progressed, since many cars dropped out of the race. During the final laps of the race, the battle to keep the lead "see-sawed" between Ray Harroun and Ralph Mulford who had the more powerful Lozier. But, because Mulford had to make more frequent pit stops for tire changes, Ray Harroun drove his Marmon at a constant pace, and went on to win the race in 6 hours, 42 minutes, and 8 seconds, averaging 74.59 mph.

The 1912 Indianapolis 500 race was even more exciting than the 1911 event, even with fewer entries. Of the 29 entries, 24 cars qualified by one-lap time trials, but the starting positions were determined by the filing of the entry chronology. The big change for 1912 was that the drivers in the lineup were driving different makes of cars from the previous year. Ralph DePalma piloted a powerful Mercedes, powered by a 583-cu.-in. displacement engine. Joe Dawson was driving a National with a 490-cu.-in. engine.

The 1912 Indianapolis 500 Mile Race started promptly at 10:00 a.m. At the drop of the green flag, Ted Tetzlaff, driving a powerful 589-cu.-in. displacement Fiat, got the jump and lead for three laps. Ralph DePalma overtook Tetzlaff, and increased his lead lap by lap. Up to the 497th mile he had smashed all records for the intermediate distances, and was expected to be the winner, as he had a lead of 11 minutes over the nearest contender. However, between turn #3 and turn #4, a half mile from the starting line, DePalma's Mercedes broke a piston and stopped. DePalma and his mechanic, Rupert Jefkins, valiantly pushed the Mercedes over the finish line, after smiling Joe Dawson was declared the winner. Thus, the proverb: "The race is not always to the fastest."

Dawson completed the 500-mile distance in 6 hours, 21 minutes, and 6 seconds, at 78.72 mph, crossing the finish line 10 minutes and 23 seconds ahead of the powerful Fiat driven by Ted Tetzlaff. DePalma was given credit for 198 laps. A very humorous incident occurred after Howard Wilcox came in ninth in a National. Ralph Mulford driving the only car on the track, asked the chief steward if he could be awarded tenth place, and be flagged off. The request was declined insisting that Mulford must complete the entire 500 miles. Mulford got back into his Knox racing car carrying hot dogs and soft drinks. With his feet planted on the dash of the car, he circled the track at 35 to 40 miles per hour, until he had accumulated 500 miles. He kept the officials at the track for 8 hours and 53 minutes, by which time it was getting dark. Needless to say the

Joe Dawson was the winner of the 1912 Indianapolis 500; (right photo) being congratulated by Fred Wagner. (Source: Indianapolis Motor Speedway Corp.)

rule "that the driver must complete 200 laps" was rescinded for the 1913 Indianapolis 500 race, when the remaining cars after the 12th place finisher would be flagged and assigned positions.

During 1917 Wheeler retired from the Indianapolis Motor Speedway and sold his stock shares to James Allison. In 1923 when the Indianapolis Motor Speedway was well on its way, Carl Fisher resigned the presidency to develop Miami Beach, Florida, and Montauk Point, Long Island, New York. Allison purchased Fisher's stock and became president of the Indianapolis Motor Speedway. Since the Allison Engineering Company was occupying most of Allison's time in 1927, he prevailed on Edward V. Rickenbacker to purchase the Indianapolis Motor Speedway. Two years later on October 24, 1929, "Black Thursday"

Ralph DePalma and Rupert Jefkins pushing the disabled Mercedes. (Source: Indianapolis Motor Speedway Corp.)

descended on Wall Street in New York. This convinced Rickenbacker and the American Automobile Association Contest Board to change the racing car regulations to permit racing car owners to use passenger car "stock block" engines up to 6 liters (366-cu.-in.) displacement. This move rescued championship automobile racing, by making racing cars affordable. Rickenbacker courageously weathered the cruel depression and in spite of the shortage of cash was able to have the turns of the track resurfaced with "Kentucky Rock" (see the Rickenbacker chapter in this volume). Then came World War II and all AAA racing was terminated.

In 1945 Wilbur Shaw was able to convince Tony Hulman to purchase the speedway from Edward V. Rickenbacker, and since then all improvements are self-evident. Even with Tony Hulman's passing in 1977, the control and fate of the Indianapolis Motor Speedway stayed within the Hulman family. The control of the future of the Indianapolis Motor Speedway is in the capable hands of Tony George, president, and Mari Hulman George, chairman. In spite of the detractive remarks by "wanna-be's," the Indianapolis Motor Speedway with its capable and talented management, will carry the IMS tradition to new heights in the 21st century.

The Legacy

Although automobile racing originated in France (the Paris to Rouen race in 1892), by the turn of the century, most industrialized areas of the world were competing to establish automobile race tracks and courses.

Many of the larger cities and population centers of the United States built race tracks and race courses during the first three decades of the 20th century. Among those were: New York, Empire City, Sheepshead Bay, Los Angeles, Beverly Hills, Oakland, Minneapolis, Twin City, Cincinnati, Chicago, Maywood, Kansas City, Savannah, Charlotte, Pittsburgh, Uniontown, Altoona, Wilkes Barre, Indianapolis, and many others. Road races were held at Wilkes Barre, Pa., Elgin, Ill., Corona, Cal., and others. But, as the real estate values appreciated and spiralled upward, the race tracks had to give way to construction development.

So why did Indianapolis not only survive, but thrive? Many attempts have been made to try to explain the course of events, but many of those explanations hinged on physical factors, such as geographical areas, population centers, natural resources, prevailing climatic conditions, etc. But there were many cities with equal, if not more advantageous areas than Indianapolis. Perhaps Indianapolis' geographical location may have had some influence. It may be well to accept that some unexplainable events somehow intervened. If you believe in that sort of predestination, or the inevitability of America's constant progress, you may reason that if the "Hoosier Pioneers" had not done it someone else may have!

The legacy of the Indianapolis 500 did not just happen with the investment of "mega-bucks." The Indianapolis Motor Speedway has had its ups and downs, along with everyone else. It took years of perseverance and experience to become one of our nation's greatest traditions.

David Parry

1912 Parry Model 52 runabout. (Source: NAHC)

Although David Parry did not get the chance to enjoy the fruits of success in automobile manufacturing, he did provide Americans the finest in horse-drawn vehicles, and later with the much-needed commercial bodies for truck chassis built by Ford, Chevrolet, Maxwell, Overland, Star, and many others. In fact, today's minivans are descendants of Parry's screen and panel truck bodies.

* * *

By the turn of the century, David M. Parry was well established as the world's largest carriage manufacturer. He also built buggies, surreys, phaetons, wagons, and carts. However, in February 1907, he became addicted to automobile manufacture through his association with the Overland Automobile Company.

Overland Automobile Company

In 1904 the Overland car was built by Claude E. Cox in the facilities of the Standard Wheel Company of Terre Haute, Indiana. Approximately 25 cars were built and sold during 1904, but they were not profitable, so Cox re-evaluated his facilities in Terre Haute, and decided that the area was too small and the machine shop inadequate. Coincidentally, Standard Wheel Company had a plant in Indianapolis that wasn't being used. So in January 1905, with the concurrence of Charles Minshall (plant manager of Standard Wheel Company), the Overland manufacturing operations were moved to the Standard Wheel plant in Indianapolis.

The Standard Wheel plant in Indianapolis happened to be near Parry Manufacturing Company, and David Parry would sometimes observe the operation of the Overland manufacture. He was fascinated by it, and became fond of Claude E. Cox and the Overland car.

On February 21, 1906, Cox entered into a contract with Standard Wheel for the purchase of the Overland machinery, material, and good will for $9,000. He also leased the building and real estate from Standard Wheel. However, Cox found himself in deep financial trouble. J.H. Keyes, president of Standard Wheel Company, demanded that Cox and Overland vacate Standard Wheel premises in 60 days.

Parry Rescues Cox

Fortunately for Cox, Standard Wheel manufactured wheels for Parry. During the course of business, Cox frequently came in contact with David Parry. Parry had experimented with horseless carriages in the 1890s, and was receptive to Cox's ideas. Cox's sincerity and enthusiasm intrigued Parry, and when he saw the newly designed, four-cylinder-engine-powered Overland car, he knew he wanted to be part of this exciting venture. So Parry offered Cox financial aid, and they agreed to the formation of a company for the purpose of building the Overland car. This was Parry's first venture into manufacturing automobiles. Forty-seven Overland cars were built in 1906, but still were not profitable.

On February 19, 1907, the Overland Automobile Company was incorporated in the state of Indiana, with David M. Parry and Claude E. Cox as incorporators and directors. Parry furnished the financial backing and temporary manufacturing space. Cox was elected president, operated the factory, and built the cars. Since Parry paid off Standard Wheel's lien against Overland, he therefore held the controlling interest in the company. Parry also had financial interest in an Indianapolis traction company, as well as several other enterprises.

A Quick Failure

While the buyers' response to the Overland car was phenomenal, the production was not; by the time of the Wall Street panic on October 22, 1907, the Overland liabilities exceeded $80,000. Working capital dried up, as Parry's enterprises were hard hit by the panic to the extent that Parry temporarily lost possession of his home due to a mortgage foreclosure procedure. With Parry unable to furnish any more operating funds, Overland production came to a halt.

Parry's creditors, believing him to be bankrupt, allowed him to keep his Overland stock, judging it to be worthless. John N. Willys worked out an arrangement with Cox and Parry whereby they would turn over two-thirds of the Overland common stock to Willys and E.B. Campbell. In return for Willys' effort to save Overland in its financial crisis, Cox and Parry would receive no payment for the stock surrendered. In the final arrangement, most of the stock surrendered came from Cox's original shares. David Parry was insistent on this, and while Cox was reluctant, he finally agreed. Cox was able to hold on to 66 shares of his original stock.

The Overland Automobile Company was reorganized with Parry and Campbell on the board of directors. Parry held on to his Overland stock until mid-1909, when he sold it to Willys for $250,000.

Parry Automobile Company

After selling his Overland stock, Parry hastily organized the Parry Automobile Company, and incorporated it on July 29, 1909. It was capitalized at $1,000,000, of which only $150,000 was paid in. Parry leased seven

1910 Parry Model 40 touring car. (Source: NAHC)

buildings from the Standard Wheel Company of Indianapolis, and within three months was employing about 390 workers and was able to ship cars to dealers.

The 1910 Parry models were powered by a four-cylinder, valve-in-head engine; power was transmitted by means of a cone clutch, sliding gear transmission, shaft drive, and a bevel gear rear axle. The suspension was by two parallel semi-elliptical springs in front, two full-elliptical springs at the rear.

	1910 Parry
Wheelbase (in.)	115
Price	$1,285-1,485
No. of Cylinders / Engine	4
Bore x Stroke (in.)	4.25 x 4.50
Horsepower	35 adv, 28.9 ALAM
Body Styles	runabout, touring car
Other Features	

Motor Car Manufacturing Company

Approximately 900 cars were built and sold during 1910; however, by overextension in spending, and only $150,000 paid in capital, Parry found himself with little working capital. On December 3, 1910, Judge Carter of the Superior Court placed the Parry Automobile Company in the hands of a receiver, the Union Trust Company of Indianapolis. Upon reorganization on January 18, 1911, David Parry was out. W.C. Teasdale was named president, G.O. Simons vice president, W.K. Bromley secretary, and F.R. Dorn treasurer. At the same time the name of the company was changed to Motor Car Manufacturing Company. The Parry car was continued through 1911 and 1912.

The New Pathfinder

On September 27, 1911, the new and larger 1912 Pathfinder was introduced. The drive was by means of a cone clutch, a four-speed sliding-gear transmission, a single Spicer universal joint, and a torque tube propeller shaft to the bevel gear rear axle differential. Suspension was by parallel semi-elliptical springs in front and rear.

	1912 Pathfinder	**1914 Pathfinder Six**
Wheelbase (in.)	120	134
Price	$2,185-2,500	$2,750
No. of Cylinders / Engine	L-4	L-6
Bore x Stroke (in.)	4.12 x 5.25	4.12 x 5.25
Horsepower	40 adv, 27.2 ALAM	40 adv, 40.9 ALAM
Body Styles	touring, armored roadster, cruiser, Martha Washington coach	Leather Stocking touring
Other Features	dual Bosch magneto ignition system	

On November 13, 1913, a new 1914 Pathfinder Six was announced. The six cylinders were cast in three's, mounted on a cast aluminum crankcase, supported by three babbited bearings. The engine was manufactured by Continental Motors Company. The Pathfinder Six featured a new V-shaped radiator core and shell, giving a streamlined effect of graceful transition from the hood and body cowl. The fuel tank was located under the cowl. The drive was on the left side, with the change-gear and brake levers centrally located. The power was transmitted by means of a cone clutch, a four-speed transmission, a single universal joint, a torque-tube enclosed propeller shaft, through bevel gears, and a full-floating axle. The four-cylinder-powered Pathfinder Series XIV, priced at $2,175 to $2,500, was continued unchanged during 1914.

1914 Pathfinder cruiser and touring car. (Source: NAHC)

The 1915 Pathfinder was announced on June 10, 1914, consisting of two lines. The larger of the two models was continued with the same specifications as the 1914 Six, but was called the "Leather Stocking" model. The smaller six was an entirely new car, identified as Model VII. The domestic sales in 1915 were estimated at 1,000 units.

	1915 Pathfinder Leather Stocking	1915 Pathfinder VII
Wheelbase (in.)	134	124
Price	$2,750-2,997	$2,220-2,322
No. of Cylinders / Engine	L-6	L-6
Bore x Stroke (in.)	4.12 x 5.25	3.75 x 5.25
Horsepower	40 adv, 40.9 ALAM	34 adv, 33.7 ALAM
Body Styles	seven-passenger touring and limousine	roadster, cruiser, Daniel Boone touring
Other Features		

Pathfinder Automobile Company

On May 24, 1915, the Motor Car Manufacturing Company was reorganized under the name of Pathfinder Automobile Company. The articles of incorporation were with the Indiana Secretary of State, with an authorized capitalization of $250,000. C.W. Richards, Leo Raminsky, and G.I. Lufkin were named as incorporators. On June 17, 1915, the Pathfinder Automobile Company revealed its plans for 1916, which contained some surprising developments, including the new twelve-cylinder-engine-powered cars.

1916 Pathfinder Twelve sport touring and cloverleaf roadster at the back of the Pathfinder plant. (Source: Indianapolis Motor Speedway Corp.)

1916 Pathfinder LaSalle touring car. (Source: NAHC)

	1916 Pathfinder Twelve	1916 Pathfinder Six
Wheelbase (in.)	130	122
Price	$2,750-4,250	$1,695
No. of Cylinders / Engine	12 (overhead camshaft)	L-6
Bore x Stroke (in.)	2.87 x 5.00	3.75 x 5.25
Horsepower	60 adv, 39.6 ALAM	50 adv, 29.4 ALAM
Body Styles	three-, four-, five-, seven-passenger LaSalle touring, cloverleaf roadster, Berline limousine	five- and seven-passenger touring, two- and four-passenger roadster
Other Features		

The power flow in the Pathfinder Twelve was through a disc clutch, sliding gear transmission, two Spicer universal joints on a hollow tube propeller shaft to a bevel gear full-floating rear axle. The thrust was taken by the semi-elliptical cantilever rear springs. The Pathfinder Six models were known as the "Fremont" models. For 1917 only the Pathfinder Twelve was built, with the same specifications and prices as the 1916 models.

In 1916, two transcontinental runs were made using Pathfinder Twelves. On April 28, 1916, Ezra Meeker, a transcontinental traveler, left Washington, D.C. on a cross-country trip in his "schoonermobile" bound to Olympia ,Washington, a distance of 3,560 miles. Meeker, who was 85 years young, had made three trips across the Oregon Trail by ox team. The schoonermobile was equipped with all the comforts dear to the hearts of the overland pioneers of the bygone generation. Signs decorating the vehicle explained the object of his journey: to retrace the Cumberland road and the Oregon Trail, and to report to Congress the condition and the probable cost of building a transcontinental military highway.

On July 3, 1916, a Pathfinder Twelve driven by Walter Weidley (son of the designer and builder of the Weidley Twelve engine which powered the car) left San Diego, California, to cross the continent locked in high gear. The car was met by Pathfinder owners and the officials of the American Automobile Association when it arrived in New York. The car was then driven to Sheepshead Bay Speedway, where it was driven over a measured course at 60 mph as the final demonstration. The car was officially sealed by the AAA representative, Al G. Waddell of Los Angeles. AAA officials and officers of the Lincoln Highway Association inspected the seals frequently over the entire route. The car had covered 4,921 miles by the completion of the test at the Sheepshead Speedway. No attempt was made to acquire a high speed average.

Ezra Meeker and his schoonermobile, 1916. (Source: Indianapolis Motor Speedway Corp.)

(Source: NAHC)

A Swift End

In spite of the favorable publicity in 1916 for the Pathfinder Twelve, World War I created material shortages which greatly affected the company. Nevertheless, as late as March 19, 1917, Pathfinder Automobile Company had expansion plans. They had consummated a deal by which they would capitalize for $5,000,000 to greatly enlarge the plant and manufacturing capacity. The A.R. Scheffer Company had underwritten all the stock, comprised of $3,000,000 in common and $2,000,000 in preferred stock, which they had intended to place on the market. All the officers were to retain their positions. But it was all for naught, because by December 1917, Pathfinder Automobile Company was in the hands of the receiver. In March 1918, all the machinery, tools, equipment, and unfinished cars, inventoried at over $500,000, were sold by the Samuel L. Winternitz and Company by auction for approximately $59,000.

Meanwhile, What Happened to Parry?

When he left Parry Automobile Company in 1911, David Parry had the wisdom to place his interest back in the Parry Manufacturing Company and to continue to build the finest buggies, surreys, wagons, and carriages. Unfortunately he died in 1915, but his brothers carried on the tradition, and even expanded the company. On November 20, 1915, the Parry Manufacturing Company engaged in the manufacture of truck and commercial vehicle bodies, tops and trailers, with specially designed couplers. They also continued to produce horse-drawn vehicles until after the Armistice on November 11, 1918.

On June 9, 1919, the Parry Manufacturing Company merged with the Martin Truck and Body Corporation of York, Pennsylvania, to form the Martin-Parry Corporation to manufacture commercial vehicle bodies and truck components. The directors were John J. Watson, chairman of the board, S.C. Parry, Guy E. Tripp, James F. Shaw, F.M. Small, Robert I. Burr, Walter R. Herrick, and Geoge R. Walbridge. On January 7, 1924, the Martin-Parry Corporation held a three-day convention of branch and district managers, at which they announced the full line of all-steel dump bodies in addition to the 38 standard types including garbage models with or without lids.

Martin-Parry built numerous special truck bodies during World War II. After the war they became the York-Hoover Corporation, and established the York Motor Express Company in York, Pennsylvania, and the York Trailer Company Limited in England, which are still in business today. The numerous trucks, tractor and trailer rigs, and most all commercial hauling was made possible by Parry, Fruehauf, and other trailer pioneers.

Hugh Chalmers

Throughout his life, Hugh Chalmers unselfishly gave of himself through financial backing and business guidance to his associates and his employees, helping them establish their own automobile manufacturing enterprises, including the Hudson Motor Car Company, the Saxon Motor Car Company, and other automobile selling organizations. Many industrial greats owe their start to him, the modest "unsung pioneer."

* * *

Hugh Chalmers was born in Dayton, Ohio, on October 3, 1873, and got his first job at the National Cash Register Company in 1887. His sales skills catapulted him through the ranks, and by the time he was 24 years old he was earning $72,000 per year as vice president in charge of sales. The "boy wonder" was able to retire at the age of 27. It was at this point in his life that he was sought out by Roy Chapin, Howard Coffin, and Edwin Thomas.

Who's Who

Roy Dikemam Chapin, born on February 23, 1880, was an engineering student at the University of Michigan. In the spring of 1901, after a disastrous fire destroyed the Olds Motor Works plant in Detroit, Chapin joined the Olds Motor Works as a gear filer, working under Jonathan D. Maxwell. Chapin was also a qualified photographer and helped with the publication of the Oldsmobile sales catalog. In late October 1901, he was selected to drive a curved-dash Oldsmobile from Detroit to New York as a publicity promotion, arriving in New York on November 5, 1901, in time for the opening of the Second National Auto Show in New York.

Roy D. Chapin. (Source: NAHC)

Edwin R. Thomas. (Source: NAHC)

Howard E. Coffin. (Source: NAHC)

Despite the fact that Chapin and the car were so covered with mud that the doorman insisted that Chapin go to the service entrance, the car was favorably displayed and excitingly received at the show.

Edwin Ross Thomas, a bicycle and motorcycle manufacturer, organized the E.R. Thomas Motor Company in Buffalo, New York, in 1900. His first cars were built in 1902, and by 1908 his famous "Thomas Flyer" had won the New York to Paris race, which began in Times Square on February 12, 1908, and ended in Paris on July 30, covering the 13,341-mile land route via California, Siberia, Russia, and Central Europe in 88 days. (See the Thomas chapter in Volume 2.)

Howard E. Coffin graduated with an engineering degree from the University of Michigan in 1902 and joined the Olds Motor Works as a mechanical engineer.

The Birth of Thomas-Detroit Company

In April 1906, after Edwin R. Thomas agreed to give his financial support to Chapin, Coffin, Frederick Bezner, and James Brady to organize the Thomas-Detroit Company, the "four horsemen" left the Olds Motor Works. In terms of organization and production, the Thomas-Detroit Company started on a good course; however, they felt that in order to be successful, they must infuse new blood and energy into the sales end of the business. Chapin and Coffin somehow successfully conspired to lure Hugh Chalmers from retirement to engage in the automobile business. Thus started the exodus from Dayton to Detroit.

General view of Thomas-Detroit assembling floor showing arrangement of chassis. (Source: NAHC)

Chalmers Takes Over

On November 16, 1907, Chapin and Coffin prevailed upon Chalmers and convinced him to buy the E.R. Thomas interest in the Thomas-Detroit Company. Thomas remained president of the E.R. Thomas Motor Company of Buffalo, New York, until 1912, when the company went into receivership. The cars carried the Thomas-Detroit name to the end of 1907, at which time the company was renamed the Chalmers-Detroit Company. The Chalmers-Detroit car was introduced in 1908 and remained in production through 1923. By 1909, Chalmers gained complete control of the company and renamed it the Chalmers Motor Company. The Chalmers-Detroit car was renamed Chalmers in 1910.

Around this time Chalmers and the others were involved in the creation of the Hudson Motor Car Company. For that complete story, see the Hudson chapter in Volume 2.

The Saxon Motor Car Company

Harry W. Ford came to Detroit from the National Cash Register Company in Dayton to join Chalmers in 1909 as advertising manager. During 1913, Hugh Chalmers gave financial backing to Ford (no relation to Henry Ford) to organize the Saxon Motor Car Company, even though Ford was Chalmers' sales manager. The Saxon Motor Car Company was incorporated on November 1, 1913, with Hugh Chalmers, Harry W. Ford, Percy Owen, and Lee Counselman as the original organizers. Chalmers encouraged Ford and gave him all needed assistance to get started, including selling the Saxon car through Chalmers dealers.

The 1909 Chalmers-Detroit.
(Source: NAHC)

1910 Chalmers. (Source: NAHC)

1913 Chalmers.
(Source: NAHC)

1911 Chalmers. (Source: NAHC)

The Saxon car was an immediate success because the Saxon runabout was graceful, economical, and beautiful to behold. The sales far outdistanced the production capabilities because of the modest price of $395. Sales and profits continued favorably through 1916; however, a disastrous fire on February 3, 1917, completely destroyed the plant. The costly building of a new plant at Wyoming Road and McGraw Avenue increased the manufacturing costs during 1917–1918, which in turn increased the selling price to such a level that the car was no longer competitive. The final blow to the Saxon Motor Car Company was the untimely death of Harry W. Ford on December 23, 1918, during the flu epidemic. Several attempts were made at reorganization, but Saxon ceased production in 1922.

Chalmers Motor Company

In the meantime Chalmers Motor Company was experiencing cash flow and other financial difficulties. Hugh Chalmers, Walter Flanders, and other officers of the Chalmers and Maxwell organizations had several conferences. The result was that on September 10, 1917, the stockholders of Chalmers Motor Company ratified a refinancing plan whereby Maxwell would advance Chalmers Motor Company $3,000,000, and Chalmers would lease the plant and assets of the Chalmers Motor Company to Maxwell Motor Company Incorporated for five years.

On September 17, 1917, Maxwell Motor Company Incorporated assumed complete control of the Chalmers Motor Company under the terms of the five-year lease arranged the previous week. The reorganization changed the name of Chalmers Motor Company to Chalmers Motor Corporation. Hugh Chalmers resigned as president but remained as board chairman. Walter Flanders and the other officers of Maxwell remained in that capacity for both corporations. Maxwell Motor Company Incorporated, after taking over the facilities of

1919 Chalmers.
(Source: NAHC)

Chalmers, continued to build Chalmers cars and used the surplus plant area to build Chalmers 3-ton trucks for commercial use and four-wheel-drive trucks for the U.S. Army.

During 1918, 80% of the capacities of Maxwell and Chalmers plants were devoted to war materiel production. This situation lasted until December 31, 1918, when the production of war materiel was curtailed and the production of cars and trucks gradually proceeded. While the production of war materiel was profitable, the return to civilian production and the subsequent depression of 1920 brought hard times for Maxwell-Chalmers and forced them into receivership in August 1920. Although the combined statements of the Maxwell Motor Company and Chalmers Motor Corporation showed the net earnings for the eleven months ending June 30, 1919, as $1,987,900, the depression of the 1920s and the loss of car sales required Maxwell-Chalmers to seek financial help from banking institutions.

However, in order to receive financial backing, the banking firms insisted on a plan to merge Maxwell and Chalmers that included the establishment of a management committee to be chaired by Walter P. Chrysler who would be in charge of the plant operations and the physical side of the merger. Chrysler took on this great responsibility in January 1921, concurrently with his responsibilities as executive vice president of the Willys-Overland Company and the Willys Corporation. By the end of 1921, Chrysler severed all his connections with the Willys organizations and devoted his full time to the Maxwell-Chalmers merger and organization. (For the full story, see the Chrysler chapter in Volume 1.)

As for Hugh Chalmers, during World War I he was designated as the representative of the automotive industry before the War Industries Board. During that period, he was also active in the Chalkis Manufacturing Company, manufacturers of anti-aircraft guns. He passed away on June 2, 1932, in Beacon, New York, following a heart attack and pneumonia complications at age 59.

For the continuation of the story of the "four horsemen," see the Hudson Motor Car Company chapter in Volume 2.

1923 Chalmers.
(Source: NAHC)

Chalmers Cars

	Wheelbase (in.)	Price	No. of Cylinders/ Engine	Bore and Stroke (in.)	hp (ALAM/NACC)
1908 Chalmers-Detroit F	110	$1,500	L-4	4.00 × 4.75	25.6
E	112	$2,750	L-4	5.00 × 4.75	40.0
1909 Chalmers-Detroit 30	110	$1,500	L-4	4.00 × 4.75	26.6
40	112	$2,750	L-4	5.00 × 4.75	40.0
1910 Chalmers 30	115	$1,500	L-4	4.00 × 4.75	25.6
40	120	$2,750	L-4	5.00 × 4.75	40.0
1911 Chalmers 30	115	$1,500	L-4	4.00 × 4.75	25.6
40	120	$2,750	L-4	5.00 × 4.75	40.0
1912 Chalmers 30	115	$1,500	L-4	4.00 × 4.75	25.6
40	120	$2,800	L-4	5.00 × 4.75	40.0
1913 Chalmers 16	115	$1,500	L-4	4.00 × 4.50	25.6
17	118	$1,950	L-4	4.25 × 5.25	28.9
18	130	$2,400	L-6	4.25 × 5.25	43.8
1914 Chalmers 14	120	$1,785	L-6	3.37 × 5.00	27.4
24	132	$2,175	T-6	4.00 × 5.50	38.4
1915 Chalmers 32	120	$1,400	L-6	3.12 × 5.00	23.4
26R	125.5	$1,650	L-6	3.50 × 5.50	29.4
M-6	132	$2,400	T-6	4.00 × 5.50	38.4
1916 Chalmers 5-15	115	$1,000	L-6	3.25 × 4.50	25.4
7-22	122	$1,350	L-6	3.25 × 4.50	25.4
1917 Chalmers 6-30	115	$1,050	L-6	3.25 × 4.50	25.4
6-30	122	$1,350	L-6	3.25 × 4.50	25.4
1918 Chalmers 6-30	117	$1,485	L-6	3.25 × 4.50	25.4
1919 Chalmers 6-30	117	$1,695	L-6	3.25 × 4.50	25.4
6-30	122	$1,795	L-6	3.25 × 4.50	25.4
1920 Chalmers 35C	117	$1,795	L-6	3.25 × 4.50	25.4
35B	122	$1,945	L-6	3.25 × 4.50	25.4
1921 Chalmers 35C	117	$1,795	L-6	3.25 × 4.50	25.4
35B	122	$1,945	L-6	3.25 × 4.50	25.4
1922 Chalmers 35C	117	$1,295	L-6	3.25 × 4.50	25.4
35B	122	$1,395	L-6	3.25 × 4.50	25.4
1923 Chalmers Y	117	$1,185	L-6	3.25 × 4.50	25.4
Y	122	$1,345	L-6	3.25 × 4.50	25.4
1924 Chalmers Y	117	$1,185	L-6	3.25 × 4.50	25.4
Y	122	$1,295	L-6	3.25 × 4.50	25.4

Harry Jewett

While the name of the company was Paige-Detroit, the name Harry M. Jewett was almost synonymous with the founding and development of the company. After earning a respected place in the fine-car market, a couple of missteps outside their niche caused the company's ultimate downfall.

* * *

Harry Mulford Jewett was born on August 14, 1870, in Elmira, New York, the son of Arthur LeRoy Jewett and Gertrude Mulford Jewett. He graduated from Notre Dame with a degree in civil engineering, and was also a prominent track star. In 1891 Harry ran under the colors of the Detroit Athletic Club and became the undisputed world champion in the 100- and 220-yard dashes.

His first occupation was as an assistant engineer on the Chicago Drainage Canal, and during 1891 and 1892 he was assistant engineer for the Michigan Central Railroad. For the next two years he managed coal mines for the W.P. Rend and Company, operators in the Hocking Valley coal fields in Ohio.

Having found partners, he struck out on his own in 1895, organizing the Jewett, Bigelow and Brooks Company, retail coal distributors in Detroit, Michigan, as well as the Jewett Phonograph Company.

On April 29, 1898, Harry M. Jewett was called to active duty, along with other members of the Michigan Naval Brigade, to serve as gun captain aboard the U.S.S. Yosemite in the Spanish-American War. His brother, Edward H. Jewett, also served aboard the U.S.S. Yosemite as gunner's mate. After the war, Harry and Edward, along with Bigelow and Brooks, became owners of the Big Sandy Coal and Coke Company and

Fred O. Paige
(Source: Burton Historical Collection)

the J.B.B. Colleries Company. Many shipmates aboard the U.S.S. Yosemite went on to become automotive giants in subsequent years.

On February 19, 1901, Harry Jewett married Mary Visscher Wendell of Detroit, and in 1903 decided to make Detroit his home. They had two children, Eleanor Osborne Jewett and Edward Huntting Jewett II.

Fred Paige

Fred O. Paige was president and general manager of the Reliance Motor Car Company, which built two-cylinder-engine-powered cars until February 1907, at which time production concentrated on motor trucks only. In early 1909 the Reliance Motor Truck Company was sold to the General Motors Company. Later that year Paige decided to go it alone and build a car bearing his name.

The financial success of their navy shipmates may have prompted the Jewett brothers to financially support Fred Paige in the organization of the Paige-Detroit Motor Car Company. Harry Jewett may have been a bit overconfident in Paige's automotive design, engineering, and manufacturing abilities, since the Reliance Motor Car Company could not be considered a raving success.

The Paige-Detroit Motor Car Company

The Paige-Detroit Motor Car Company was organized on September 28, 1909, with Fred Paige as president, and financial support from Harry and Edward Jewett. At this time the Jewett brothers were still president and vice president of the Jewett, Bigelow and Brooks Incorporated, coal distributors in Detroit.

The first Paige car was built in 1909 in a plant at 245-255 Twenty First Street at the southwest corner of Lafayette Avenue in Detroit, Michigan. The car had a three-cylinder, two-stroke-cycle engine, which proved to be anything but satisfactory. The choice of this small three-cylinder engine may have been influenced by Paige's experience at Reliance. The car and the company did not make the favorable impression that the organizers and financial supporters had expected it would.

On February 1, 1910, the Paige-Detroit Company was reorganized with Harry Jewett as president and Edward Jewett as vice president. Fred Paige left the company. The capital stock of the company was increased from $100,000 to $250,000. Jewett and his associates: Gilbert W. Lee, grocer and banker; William B. Cady, paper manufacturer; Charles B. Warren, counsel; Charles F. Hodges, manufacturer and banker; Willis E. Buhl, foundry owner; Edward D. Stair, publisher; and Sherman L. DePew, shoe manufacturer, took hold of the company with the idea of designing, building, and marketing an automobile according to the finest scientific engineering, manufacturing, and business methods. To ensure success, the 1910 Paige-Detroit car was

The first Paige three-cylinder engine, 1909. (Source: NAHC)

redesigned using a four-cylinder, four-stroke-cycle, water-cooled engine, along with many other improvements. The car was priced at $800 and Detroit was dropped from its name.

Although the first attempt to manufacture cars in 1909 was a dismal failure, the second year the company produced about 300 cars. In August 1910, Jewett persuaded James F. Bourquin, Chalmers Motor Company superintendent, to leave Chalmers and join Paige-Detroit as general factory manager.

1910 Paige (Source: NAHC)

The Paige-Detroit for 1911
Fully guaranteed for one year
4-Cycle—4-Cylinder
Full 25 H.P. Motor
$800
A Real Opportunity for Dealers
The dealer who is so fortunate as to have the Paige Agency for 1911 is bound to make customers, friends and money. We feel justified in saying that the 1911 Paige is the first car of the year that is distinctly without competition. While it has no new or undeveloped engineering features, still it is far from conventional, and possesses many distinctive advantages and selling points which at once appeal to the dealer and the buyer, Mail the coupon below today—we will send you complete catalog and Agency Proposition by mail.
Some Features We Must Mention
even if we have to crowd the type. This is the only car having a three-point spring suspension with full elliptic spring in rear. This insures easy riding and no tilting. No person is too big for this car. The motor is of our own design and permits a quick "get away" and shows the way to a lot of expensive speeders, developing speed from 5 to 55 miles per hour. It has been thoroughly tested on hills as well as in mud and sand, showing extraordinary endurance. Its tires are above the average for this type of car, being 32x3½ inches. It has the highest developed oiling system with sight feed on dash, the same as in the high-priced cars. It is equipped with a 10-inch brake, as well as a disc clutch. Our Touring Car has a detachable tonneau which, with the rear seats off, has a carrying capacity of 36x31 inches. This feature makes it most desirable for light delivery work.
4 Passenger Touring Car (with detachable tonneau)
$900
SEND TO-DAY
Paige-Detroit Motor Car Co., 245-255 Twenty-First St., Detroit, Mich.
Paige-Detroit Motor Car Company,
245-255 Twenty-First Street,
Detroit, Michigan.
Gentlemen:
Kindly send me by an early mail photographs and specifications of the Paige-Detroit, your catalog and your agency proposition.
Name....................................
Address................................
A Word About Our Guarantee
Protects Buyer and Seller
We guarantee our car for one year. This is a definite guarantee, nothing misleading or contradictory. We will furnish a new part for any part found defective in material and workmanship if the old part is returned to our factory for inspection, freight prepaid. This refers to all parts of the car, but not accessories and tires. This guarantee is for one year from the date of the purchase of the car and is made to protect both dealer and buyer against imperfections which may be found even in the highest priced cars.
CATALOG READY
We have prepared a book telling the whole story and it is one you will be glad to get, for it shows photographs and gives complete specifications of all the models of this particular car which is the surprise of the year. Write us today for your copy.

Harry Jewett simply applied the wisdom and experience of other lines of industry, and introduced modern scientific business methods into motor car building, in a company where such methods were needed. His reputation as an organizer and leader was well known throughout the industry.

In December 1913, Paige-Detroit Motor Car Company moved into their new factory at the corner of Fort and McKinstry Streets in Detroit, Michigan. During the third quarter of 1913, the company enjoyed a net increase of $1,200,000. Paige-Detroit built and sold approximately 5,000 cars in 1914.

In December 1914 Paige introduced their first six-cylinder-engine-powered car, a 3.50 x 5.25-in. bore and stroke en-bloc L-head design, and during 1915 they built both four- and six-cylinder-engine-powered cars. They built and sold almost 8,000 cars in 1915, and at the end of the year paid cash dividends amounting to 68%. On August 2, 1915, Paige-Detroit Motor Car Company increased its capital to $1,000,000.

Starting in 1916 Paige-Detroit built only six-cylinder-engine-powered cars. The cars were beautifully styled, and at times Paige dared to claim it to be "the most beautiful car in America." During World War I they built a modest number of cars, along with war materiel for the United States Government. Sales were good right through 1920, from more than 12,000 cars in 1916 to approximately 15,750 cars in 1919 and 16,000 cars in 1920.

However, during 1921 Paige felt the pinch of the post-WWI depression as sales dropped to just above 8,000 cars. In an attempt to bolster sales in 1921, Paige had Ralph Mulford drive a modified "6-66" roadster with the top down at Daytona Beach, Florida, at 102.83 mph to establish a one-mile straightaway record for stock cars. For 1922 this roadster was called the "Daytona." Despite these sales efforts, Jewett felt there was a need for a medium-priced car, since the market for cars in the higher $1,750 to $3,750 price bracket was limited and there were many competitive makes.

The designers were hard at work all during 1921, and by December Paige-Detroit Motor Car Company was ready to announce a new Jewett car at the January 1922 National Auto Show in New York. The new Jewett car was powered by a Paige-built six-cylinder engine developing 50 hp in a 115-in. wheelbase chassis. While the Jewett organization was called Jewett Motors Incorporated, a subsidiary of Paige-Detroit Motor Car Company, the car was built in the Paige factory and merchandised by the Paige dealer organization. The car was moderately priced at $1,065 to $1,395 and it was an immediate success. In 1922 Paige-Detroit Motor

1918 Paige. (Source: NAHC)

Car Company built and sold 20,420 Jewett cars and 9,323 Paige cars. The 1922 sales volume amounted to $32,749,666, the net profit was $2,103,230, which provided a 12% cash dividend and a 100% stock dividend.

Also in 1922 a new engine plant was built on West Warren Avenue near Wyoming Avenue in Detroit. The new Jewett car production was moved to this plant after an addition of 500,000 sq. ft. of plant area. Later all car assembly was concentrated in this plant, while the general offices remained at the Fort and McKinstry location. On April 10, 1923, Paige terminated all commercial vehicle production due to car demand.

To further attract customers, in November 1923 Paige and Jewett adopted a weekly payment plan. Sales increased considerably, and the engine manufacturing facilities could not keep pace, so during 1924, 1925, and 1926, the Paige six-cylinder engines were built by Continental Motors Corporation while the Jewett engines were built in the Paige plants.

For 1925 Jewett added a Deluxe Brougham, a four-door model priced at $1,525. In 1925 all Paige and Jewett cars had Paige-Lockheed hydraulic four-wheel brakes. Jewett coach sales were brisk, since it was priced at $1,245, $75 under the price of the touring car. On July 8, 1925, Paige announced a new five-passenger sedan model priced at $2,345, the lowest-priced model ever built on a 131-in. wheelbase Paige chassis.

On July 22, 1925, the new executive office building on West Warren Avenue was completed, and all sales and administration executives moved there. A new engineering building was built between the existing factory buildings and the new executive offices. The factory service division offices remained at the Fort and McKinstry location.

Beginning of the End

During the first five months of 1925, Paige-Detroit Motor Car Company earned $1,682,498, a new record earning. This may have been the spark that prompted Paige-Detroit Motor Car Company to make the next momentous decision for 1926, a decision that may have been the first step toward the demise of the Paige-Detroit Motor Car Company.

In December 1925 the company announced with much fanfare the "New Day" Jewett, a two-door sedan model priced at $995. It was built on a 109-in. wheelbase chassis, powered by a 40 hp, Continental-built, six-cylinder engine. Wood construction in the body was eliminated and replaced by an all-steel body built by the Murray Corporation, which featured less-bulky corner pillars, doorposts, and window frames. The New Day Jewett was downsized in all respects: wheelbase length, engine, transmission, wheel and tire sizes. Even the hydraulic brakes were smaller.

Promotional copy to attract dealers stated:

> The new low price of this New-Day Jewett Sedan alone insures for Jewett dealers opportunities for volume business and volume profits such as they have never enjoyed before. But price is not the only appeal, nor the primary appeal, of this remarkable new car. For the new-Day Jewett Sedan is not only a new motor car, but an entirely *new type* of motor car. Built expressly to meet and master the trying and exacting demands of new-day traffic conditions—this car will find an immediate and eager market among those who have long wanted and long awaited a truly superfine small car. It is the easiest car to drive, and the safest, on the market today. Paige-hydraulic 4-Wheel Brakes, unbelievably swift acceleration, power to spare, balloon-tired comfort, an almost complete elimination of the deadly "blind-spot"—these are but a few of the features that are certain to establish new sales records for those fortunate dealers everywhere who own, or are successful in obtaining, the Paige-Jewett franchise.

1926 New Day Jewett.
(Source: NAHC)

While the New Day Jewett may have found a place in today's "econo-box" market, the former Jewett buyers were displeased and stayed away in droves. The former full-sized Jewett built through 1925 became the Paige Model 24/26 for 1926, and was available in 115- and 125-in. wheelbase lengths. Paige bodies continued to be built of wood and steel by the Briggs Manufacturing Company.

Needless to say the sales of Jewett tumbled from 28,621 registrations in 1925 to 14,560 in 1926. The Paige model 24/26 helped Paige sales somewhat; however, the total registrations of Jewett and Paige combined were 30,324 in 1926 compared to 32,324 in 1925. Meanwhile the rest of the automobile industry was enjoying a boom year with a 10% gain over 1925.

On January 4, 1927, at a board of directors meeting, W.A. Wheeler was elected president of Paige-Detroit Motor Car Company, succeeding Harry Jewett who was elected board chairman. Wheeler had been identified with Paige-Detroit organization for 14 years as head of the manufacturing division.

In a last valiant attempt to regain lost prestige in the fine car field, Paige-Detroit Motor Car Company introduced the new Paige "Straightaway Eight" at the National Auto Show in New York in January 1927. This fine car featured an eight-in-line engine having a 3.25-in. bore and a 4.50-in. stroke, rated at 85 hp. This engine built by Lycoming Manufacturing Company was coupled to a new "Warner Hy-Flex" four-speed transmission, featuring an internal tooth third-speed gear. The chassis had a 131.5-in. wheelbase, on which were mounted superbly crafted bodies. The interiors had walnut-finish instrument panels and window garnish moldings. Leather upholstery was used for open models, and the closed models were trimmed in gray velvet mohair cloth. The price range for the Paige Straightaway Eight was $2,295 to $2,795.

While the introduction of the Paige Straightaway Eight was commendable, and it could hold its own in the price bracket, the next inglorious move sealed the fate of Paige-Detroit Motor Car Company. Heaping insult upon injury to potential Paige buyers, the New Day Jewett Model 6-45 was thereafter known as the Paige 6-45.

The dollar sales volume for the first quarter ending March 31, 1927, amounted to $5,952,669 compared to $17,399,927 in the same quarter a year before. This resulted in a net loss for the quarter of $185,798 compared to a profit of $505,369 for the same period in 1926. The total Paige registrations for 1927 amounted to a dismal 18,256 cars.

Salvation came to the Paige-Detroit Motor Car Company on May 5, 1927, when the Graham brothers purchased Harry Jewett's interest in the company for a reported $4,000,000. The name of the company was changed to Graham-Paige Motors Corporation on January 1, 1928. The new Graham-Paige, the product of the Graham Brothers' ownership, was introduced at the New York National Auto Show in January 1928. In the final accounting of the Paige-Detroit Motor Car Company on February 29, 1928, the report showed a net loss of $4,643,351.

The Graham-Paige Motors Corporation history continues in the Graham brothers chapter in Volume 2.

With the sale of Paige-Detroit Motor Car Company, Harry and Edward Jewett returned to their coal mining and retail interests. During the Depression Harry established and became president of the Colonial Laundry Company. On June 15, 1933, Harry Jewett died of a heart attack at his home on Lake Shore Road in Grosse Pointe Shores, as he was about to leave for his office.

Frederick Chandler

Some companies are doomed from the start; others grow into empires. And some-times a company fills a need for a decade or so, earning handsome profits for its shareholders, until its time is up and it quietly disappears into history. This is the story of one of those.

* * *

An Early Start

Frederick C. Chandler was born in Cleveland, Ohio, on July 12, 1874. After two years of high school, at 16, he started to work for H.A. Lozier and Company of Cleveland, a manufacturer of sewing machines and bicycles. The youthful Chandler promptly earned a promotion to shipping clerk, and soon afterward was promoted to sales. By 1899, he was manager of the European headquarters and Lozier branches in London and Berlin.

In late 1899, Lozier sold four of its bicycle manufacturing plants to the American Bicycle Company, who merchandised the gasoline-engine-powered tricycle (designed by Lozier engineers) as the "Cleveland Three Wheeler." Chandler moved to Plattsburgh, New York, where Lozier continued to manufacture marine en-gines and launches. Henry A. Lozier, Sr., died in 1903, at which time Henry A. Lozier, Jr., became president of the company.

Lozier Motor Company factory works at Plattsburgh, NY. (Source: NAHC)

During 1903 and 1904, Lozier engineers were busy developing a gasoline-engine-powered automobile, which made its debut at the National Auto Show in New York in January 1905. The Lozier car was an immediate success as it gained the reputation for high quality and performance. Chandler was made manager of Lozier's sales agencies in United States and of its foreign (overseas) department. In 1910, when Lozier moved its headquarters and main plant to Detroit, Michigan, Chandler was elected vice president in charge of sales, and in 1911 he became general manager.

By 1910, Lozier was manufacturing expensive luxurious cars ranging from $3,500 to $7,000. Chandler recognized that the extremely high price would limit the sales potential. He and other Lozier executives tried to convince H.A. Lozier, Jr., of the need for a medium-priced car in the model lineup, but to no avail. So in January 1913, Chandler and four other Lozier executives resigned; the others were: Charles A. Emise, sales manager; John V. Whitbeck, experimental engineer; Samuel Regar, treasurer; and W.S.M. Mead, New York branch manager.

H.A. Lozier, Jr.
(Source: NAHC)

Chandler Motor Car Company

In February 1913, Chandler and his associates organized the Chandler Motor Car Company and incorporated it under the laws of the state of Ohio, with an authorized capital of $425,000. Temporary headquarters were set up at 925 Woodward Avenue, Detroit, Michigan, and a six-acre site was purchased on East 131st Street near the Belt Line Railroad in Cleveland, Ohio, to build the manufacturing plant.

The first Chandler prototype model was built in a small machine shop on East 65th Street in Cleveland and made its debut at the Chicago Automobile Show in February 1913. Whitbeck's creation, the "Chandler Light Six," was a huge success at the show. A great number of distributors placed large "paid for" orders for cars five months before production and shipments started. This could have been expected because the new Chandler embodied many impressive features and engineering innovations, including a powerful six-cylinder engine of 3-3/8 in. bore and 5-in. stroke, a multiple dry disc clutch, Westinghouse starting, lighting, and ignition, and 120-in. wheelbase chassis with semi-elliptical springs in front and three-quarter elliptical rear springs. The styling was appealing and in good taste, with a low flat hood hinged in the middle similar to that of the Lozier, Winton, and other expensive cars. The body work was well-executed and superbly finished. Then there was the unbelievable price of $1,785 for such a powerful car. No wonder sales climbed steadily and production had a difficult time keeping pace!

	1914 Chandler 14	1915 Chandler 15
Wheelbase (in.)	120	120
Price	$1,785	$1,595
No. of Cylinders / Engine	L-6	L-6
Bore x Stroke (in.)	3.37 × 5.00	3.37 × 5.00
Horsepower	27.3 NACC	27.3 NACC
Body Styles	3	6
Other Features		

	1916 Chandler 16	1917 Chandler 17	1918 Chandler 17
Wheelbase (in.)	120	123	123
Price	$1,295	$1,395	$1,675
No. of Cylinders / Engine	L-6	L-6	L-6
Bore x Stroke (in.)	3.37 × 5.00	3.50 × 5.00	3.50 × 5.00
Horsepower	27.3 NACC	29.4 NACC	29.4 NACC
Body Styles	7	5	7
Other Features			

In 1914, Chandler reduced the selling price from $1,785 to $1,595. While many competitive automobile companies were engaged in racing, hill climbs, and other forms of competition, Chandler chose to emphasize economy in purchase price and cost of operation. During May 1914, the testing laboratory of the Automobile Club of America conducted four certified fuel economy tests, using a 1914 Chandler touring car on Long Island and in New York City's Central Park. The tests covered 86.5 miles, and the Chandler vehicle had a gross weight of 3725 lb. including four passengers. The average distance traveled per gallon of gasoline consumed was 21.6 miles, with the greatest economy in the fourth run at 23.7 miles.

Approximately 550 Chandler cars were manufactured and sold in 1913, and in 1914 the sales increased to almost 2,000 units. In 1915, the price was reduced to $1,295, and production and sales reached such a high (about 7,000 cars) that they needed to expand production facilities. In December 1915, Chandler Motor Car Company was reorganized and capitalized at $7,000,000. Production and sales for 1916 doubled that of 1915.

The Chandler plant in 1922. (Source: P.M.C. Co.)

After the United States entered World War I, the availability of material declined, causing a drop in automobile production and sales. Chandler management and engineers recognized the need for a lower-priced car and developed such a vehicle during 1918.

Expanding with the Cleveland Automobile Company

In 1919 Chandler and associates incorporated the Cleveland Automobile Company under the laws of the state of Delaware, capitalized at $1,400,000. The officers of the new company included: Frederick C. Chandler, president; Samuel Regar, treasurer; John V. Whitbeck, chief engineer; and other Chandler officials. The car known as the Cleveland Six was built in a separate plant, completed in August 1919. The Cleveland Six, built on a 112-in. wheelbase chassis, featured a six-cylinder valve-in-head engine.

	1919 Chandler New Six	1919 Cleveland Six
Wheelbase (in.)	123	112
Price	$1,795	$1,385
No. of Cylinders / Engine	L-6	IH-6
Bore x Stroke (in.)	3.50 × 5.00	3.00 × 4.50
Horsepower	29.4 NACC	21.6 NACC
Body Styles	8	roadster, touring, coupe, sedan
Other Features		

The Cleveland Six was modestly priced and was initially merchandised through independent Cleveland dealers, who soon gave way to Chandler-Cleveland dual dealerships. By the end of 1920, Chandler-Cleveland boasted more than 1,000 dealers. While 1919 was a slow year for Chandler-Cleveland as the post-WWI depression gripped the United States, in 1920 they produced and sold more than 2,300 cars, realizing a gross income of over $4,000,000 and a net profit of approximately $1,700,000.

	1920 Chandler Six	1920 Cleveland 40
Wheelbase (in.)	123	112
Price	$1,895	$1,385-2,195
No. of Cylinders / Engine	L-6	IH-6
Bore x Stroke (in.)	3.50 × 5.00	3.00 × 4.50
Horsepower	29.4 NACC	45 adv, 21.6 NACC
Body Styles	6	roadster, touring, coupe, sedan
Other Features		

The year 1921 was a time of soaring inflation and economic depression. Chandler-Cleveland, caught in this squeeze, raised the prices of the Cleveland cars. However, they soon realized that the price increase was counterproductive, and in June 1921, reduced them again. Meanwhile, Chandler had introduced a two-passenger roadster for which they reduced the price as well. Further reflecting the business climate in 1921, Chandler-Cleveland production and sales dropped to a combined total of approximately 10,000 units, and the net profit was reduced to a mere $41,000.

	1921 Chandler Six	1921 Cleveland 40
Wheelbase (in.)	123	112
Price	$1,895-2,995	$1,435-2,445
No. of Cylinders / Engine	L-6	IH-6
Bore x Stroke (in.)	3.50 × 5.00	3.00 × 4.50
Horsepower	29.4 NACC	45 adv, 21.6 NACC
Body Styles	7	4
Other Features		

During 1922, Chandler-Cleveland followed a program of strict conservatism, but not at the expense of progress. Chandler and Cleveland introduced new individual fenders and step plates, eliminating the need for running boards on open models, setting a trend for the industry. The cars featured new and sportier top lines, handsome closed bodies, and drum-shaped headlamps and cowl parking lamps. Wire-spoke wheels were available as an option. The rear windows of the roadster and touring car models had the same distinctive shape as the badge (nameplate) on the radiator shell of each car line. Chandler also added a sporty five-passenger touring car to its model lineup.

Because of the handsome styling, innovative engineering features, and competitive pricing, Chandler and Cleveland enjoyed a profitable business year in 1922, manufacturing and selling a combined total of over 20,000 cars. The resulting financial picture was equally favorable. During the 1922 calendar year, Chandler and Cleveland accumulated a total income of $3,995,788 and a net profit of $1,705,788, equal to $6.09 earned on each of the 280,000 shares of common stock.

	1922 Chandler Six	1922 Cleveland 41
Wheelbase (in.)	123	112
Price	$1,595-2,395	$1,295-2,295
No. of Cylinders / Engine	L-6	IH-6
Bore x Stroke (in.)	3.50 × 5.00	3.06 × 4.50
Horsepower	29.4 NACC	45 adv, 22.5 NACC
Body Styles	new five-passenger touring	
Other Features	individual fenders and step plates	individual fenders and step plates

In January 1923, Chandler introduced the "Pike's Peak Motor," so named because the pre-production engine tests were performed on the winding steep grades of Pike's Peak in the Rocky Mountains. The six cylinders were cast en-bloc, with the removable cylinder head of L-shaped combustion chamber configuration. The camshaft, generator, and accessory drives were by a single silent timing chain.

	1923 Chandler SS-20	**1923 Cleveland 42**
Wheelbase (in.)	123	112.5
Price	$1,495-2,895	$995-1,495
No. of Cylinders / Engine	L-6	IH-6
Bore x Stroke (in.)	3.50 × 5.00	3.06 × 4.50
Horsepower	55 adv, 29.4 NACC	50 adv, 22.5 NACC
Body Styles	new seven-passenger limousine, new seven-passenger town car, and new Chummy Sedan	
Other Features	running boards	

Running boards were reinstated on the 1923 Chandler closed models. The Cleveland car prices were reduced yet again by $100. The year 1923 was the most profitable for Chandler and Cleveland, as the total combined production and sales amounted to approximately 25,000 units. Their combined income for 1923 was also a record, earning $4,041,373 with a net profit of $2,055,267, equivalent to $7.34 per share of common stock.

The year 1924 marked the introduction of Chandler's remarkable "Traffic Transmission," the absolute forerunner of all constant mesh transmissions. Built under Campbell patents, it employed all forward gears in constant mesh, and permitted easy, quiet shifting between all forward gears without clashing, regardless of vehicle speed. This was accomplished by synchronized engaging devices between the main shaft and the speed gears.

	1924 Chandler 33	**1924 Cleveland 42**
Wheelbase (in.)	123	112.5
Price	$1,485-2,995	$1,045-1,645
No. of Cylinders / Engine	L-6	IH-6
Bore x Stroke (in.)	3.50 × 5.00	3.06 × 4.50
Horsepower	55 adv, 29.4 NACC	50 adv, 22.5 NACC
Body Styles	9	10
Other Features	full running boards	full running boards

During 1924, all Chandler and Cleveland models returned to providing full running boards. The Chandler sport models featured nickel-plated radiator shells, while the radiator shell on the remainder of the models was finished in black enamel. The advantages of the Pike's Peak Motor and the Traffic Transmission were obvious to the buying public, and with the proliferation of body styles, Chandler sales were good, approximately 10,700 cars, but their profit for 1924 dropped to below that of 1923.

The 1925 models of Chandler and Cleveland were introduced in the fall of 1924 and featured balloon tires as standard equipment, and four-wheel mechanical brakes as an option. The 1925 Cleveland also featured a "one-shot" chassis lubrication system by Bowen, which lubricated the entire chassis by one pump of the system foot pedal. The Cleveland bodies were built by the Mullins Body Corporation of Salem, Ohio, who manufactured steel body parts for many other car manufacturers.

MULLINS
S T E E L   B O D Y   P A R T S

From Blueprints to Beauty

From blueprints to finished beauty is a far cry in automobile manufacturing.

Brains and skill must be supplemented by experience and modern manufacturing facilities if the designers' plans are to be faithfully embodied in the completed car.

The graceful symmetry of the new Cleveland is no less a tribute to its body parts made by Mullins, than to the artistic ability of the designer.

MULLINS BODY CORPORATION
SALEM, OHIO

Steel body parts for the automotive trade including Chrysler, Cleveland, Cunningham, Franklin, Hupmobile, Jewett, Jordan, Lincoln, Locomobile, Marmon, Maxwell, Nash, Packard, Peerless, Pierce-Arrow, Reo, Rickenbacker, Stearns-Knight, Sterling-Knight, Wills Sainte Claire, Willys-Knight.

	1925 Chandler 33A	1925 Cleveland 43
Wheelbase (in.)	123	115
Price	$1,595-2,195	$1,095-1,625
No. of Cylinders / Engine	L-6	IH-6
Bore x Stroke (in.)	3.50 × 5.00	3.12 × 4.75
Horsepower	55 adv, 29.4 NACC	55 adv, 23.4 NACC
Body Styles	new four-passenger Comrade roadster with rumble seat	
Other Features	balloon tires, optional four-wheel mechanical brakes	balloon tires, optional four-wheel mechanical brakes, Bowen lubrication system

In an attempt to motivate additional sales, Chandler and Cleveland engaged famed racing driver Ralph Mulford. In January 1925 he drove a Cleveland Six with a special rear axle ratio on the Culver City, California, race track for a distance of 1,000 miles in 12 hours, 25 minutes, and 28-2/5 seconds. A few days later on the same track, he drove a stock Chandler car equipped with a special gear ratio, at an average speed of 86.96 mph covering the 1,000 miles in 11 hours, 29 minutes, and 54 seconds. He beat the standing record by 28 minutes and 3/8 second, which had been set in 1922 by a non-stock car. On September 7, 1925, race driver Charlie Myers blazed his way up Pike's Peak 12-1/2 miles in 17 minutes and 48-2/5 seconds. His Chandler race car smashed the previous record by 26-3/5 seconds at the seventh annual hill climb world championship. Glen Schultz in a Stutz was second at 18 minutes 54-4/5 seconds.

	1926 Chandler 35	1926 Cleveland 31
Wheelbase (in.)	124	108.5
Price	$1,490-1,895	$895-1,090
No. of Cylinders / Engine	L-6	IH-6
Bore x Stroke (in.)	3.50 × 5.00	3.00 × 4.25
Horsepower	55 adv, 29.4 NACC	45 adv, 21.6 NACC
Body Styles	new seven-passenger Twentieth Century Sedan	4
Other Features	Bowen lubrication system	

Chandler cut the prices on the 1926 models and added many new improvements, including the Bowen one-shot, chassis lubrication system. The wheelbase was increased and the Pike's Peak engine featured a twin-flywheel with a single disc clutch. Three vertical bars were added to the front of the nickel-plated radiator shell, giving it a distinctive appearance. A new seven-passenger sedan (called the "Twentieth Century Sedan") was added to the line, featuring a sliding glass partition behind the front seat, making it also a limousine model.

The Merger

On December 21, 1925, a charter was filed in Dover, Delaware, for the Chandler-Cleveland Motors Corporation. This represented a merger of the Chandler Motor Car Company and the Cleveland Automobile Company, which the stockholders voted on and approved on January 15, 1926. The formal merger took place in March 1926, at which time the assets were listed at $3,000,000 and the liabilities at $1,461,000. During 1926, Chandler built and sold approximately 11,192 cars, and Cleveland approximately 9,073 cars. Chandler-Cleveland Motors Corporation posted a net profit of $401,329.

	1927 Chandler Standard 6	1927 Chandler Special 6
Wheelbase (in.)	108.5	115
Price	$995	$1,345
No. of Cylinders / Engine	L-6	L-6
Bore x Stroke (in.)	3.00 × 4.25	3.12 × 4.75
Horsepower	50 adv, 21.6 NACC	65 adv, 23.4 NACC
Body Styles	7	5
Other Features		

	1927 Chandler Big 6	1927 Chandler Royal 8
Wheelbase (in.)	124	124
Price	$1,695	$1,995-2,295
No. of Cylinders / Engine	L-6	L-8
Bore x Stroke (in.)	3.50 × 5.00	3.19 × 4.75
Horsepower	75 adv, 29.4 NACC	80 adv, 32.5 NACC
Body Styles	8	five-passenger sedan, seven-passenger sedan four-passenger coupe, roadster
Other Features		integral sun visor, velvet mohair, overstuffed cushions

The 1927 Chandler models were introduced in the fall of 1926; the Cleveland Six was discontinued and replaced with the Chandler Standard Six; the lineup also included the Chandler Special Six. January 1927 also marked the entry of the Chandler-Cleveland Motors Corporation into the eight-cylinder-engine car field: the "Chandler Royal Straight Eight" was introduced at the 1927 National Auto Show in New York City. The production and sales of 1927 Chandler cars amounted to 18,445 units. In spite of the brisk sales, Chandler-Cleveland Motors Corporation sustained a net loss of $473,109 for 1927.

The 1928 Chandler line included a number of custom-built models with lower body lines. The fenders were fully crowned, and the radiator shell was gracefully rounded and featured a new Chandler badge. Prices were reduced by $100 to $200 on almost all models. The Chandler Standard Six was identified as Model 65.

	1928 Chandler 65	1928 Chandler Special 6	1928 Chandler Big 6
Wheelbase (in.)	109	115	124
Price	$895-995	$1,135-1,235	$1,525-1,725
No. of Cylinders / Engine	L-6	L-6	L-6
Bore x Stroke (in.)	3.12 × 4.25	3.12 × 4.75	3.75 × 5.00
Horsepower	55 adv, 23.4 NACC	65 adv, 23.4 NACC	83 adv, 33.7 NACC
Body Styles	7	7	7
Other Features			

	1929 Chandler 65	1929 Chandler Big 6
Wheelbase (in.)	109	124
Price	$875-1,075	$1,525-2,025
No. of Cylinders / Engine	L-6	L-6
Bore x Stroke (in.)	3.12 × 4.25	3.75 × 5.00
Horsepower	55 adv, 23.4 NACC	83 adv, 33.7 NACC
Body Styles	7	7
Other Features		

1929 Chandler 65 speedster. (Source: P.M.C. Co.)

	1929 Chandler Royal 75	1929 Chandler Royal 85
Wheelbase (in.)	118	124
Price	$1,275-1,495	$1,795-2,295
No. of Cylinders / Engine	L-8	L-8
Bore x Stroke (in.)	3.00 × 4.50	3.37 × 4.75
Horsepower	80 adv, 28.8 NACC	95 adv, 36.4 NACC
Body Styles	5	8
Other Features		

On July 21, 1928, Chandler announced its new 1929 models, which included many refinements over the 1928 models. A new, low-priced, eight-cylinder-engine model "Royal 75" was added to the line, and the Special Six model was discontinued.

Acquisition by Hupp

Despite lower prices and the many new features, including Westinghouse power brakes, Chandler sales continued to tumble, and it was obvious Chandler-Cleveland would sustain a great loss for 1928. Thus, in November 1928, Chandler negotiated a deal with the Hupp Motor Car Company, whereby Hupp acquired control of the Chandler-Cleveland Motors Corporation. The deal was consummated on December 1, 1928. The manufacture of Chandler cars continued until May 1, 1929, building and selling approximately 7,499 cars.

After the acquisition of Chandler by the Hupp Motor Car Company, Chandler joined his son, Frederick, Jr., in the Chandler Products Company. Frederick, Jr., who had been president, died in 1943, at which time the senior Chandler assumed the presidency and continued operating the company until shortly before his death in 1945.

The Hupp interest in the Chandler properties was in the manufacturing facilities. Hupp Motor Car Company built the low-priced 1930 Hupmobile Six in the Chandler-Cleveland plants. After the demise of Hupp Motor Car Company in 1940, some of the Cleveland plants were sold to the Weatherhead Company, manufacturers of hydraulic hoses and other components for the automotive industry.

Edward Rickenbacker

From a humble stonecutter at $1 per day to automobile racing driver, from chauffeur for Col. Billy Mitchell to hero and "Ace of Aces" aircraft pilot during WWI, and to an automobile bearing his name, Edward Rickenbacker had a profound influence on aviation, automobile racing and the Indianapolis Motor Speedway, and airline development.

* * *

Early Racing Days

Edward V. Rickenbacker was born on October 8, 1890, in Columbus, Ohio. At 14 he worked as a monument stonecutter at $1 per day. His first automobile racing experience came as a riding mechanic with Lee Frayer in the Vanderbilt Trophy Race on September 22, 1906. He then started driving and won his first race at Red Oak, Iowa, in 1909. He won a 100-mile race at Columbus, Ohio, and won nine out of ten races at county fairs in Nebraska and Iowa during 1910 (none of these races were recorded because AAA did not keep records until 1916).

His history at Indianapolis Motor Speedway included driving relief for Lee Frayer in 1911, being forced out of the race on lap 40 in 1912, and finishing in tenth place in a Duesenberg in 1914. Rickenbacker established a world record in a Blitzen Benz at Daytona in 1914 at 134 mph. He headed up the Maxwell racing team in 1915 and 1916, and won seven championship races from 1914 to 1916. He introduced a steel racing helmet at Indianapolis in 1916, and innovated cowl strengthening by reinforcing it with a steel wheel rim.

The 1916 Indianapolis 500 marked the first use of steel helmets. Right to left: pole position #18 Aitken, #5 Rickenbacker, #28 Anderson, and #17 Resta. (Source: Indianapolis Motor Speedway Corp.)

The War Years

In January 1917 Rickenbacker attempted to organize the Aero-Reserves of America with drivers DePalma, Harroun, Mulford, Cooper, and others, but could not convince U.S. Army officials. He joined the Army in early 1917 and went to France as chauffeur for Colonel Billy Mitchell. When war was declared in 1917, the United States Army Signal Corps had only twelve combat planes, two training planes, and eleven planes on order.

Rickenbacker was assigned as engineering officer at the flying school in Tours, France. After 25 hours of flying time in 17 days he became a pilot and was commissioned First Lieutenant in the U.S. Army Signal Corps. He was assigned to a new Spad (French aircraft) in the "Hat in the Ring" Squadron #94. His first air victory was on April 29, 1918, but because of a mastoiditis infection he had no flying time during June and July. On September 24, 1918, he was appointed commander of Squadron #94. He downed 14 enemy aircraft during October, and on October 30th he got his 25th and 26th victories. Squadron #94 destroyed 69 enemy planes during that period. He received 12 medals, awards, and citations during 1918 alone. After the Armistice on November 11, 1918, Captain Edward V. Rickenbacker enjoyed the respect, popularity, and admiration of all the American people. He was "America's Hero, the Great American Aviation Ace of World War I."

Breaking Into the Car Business

Although Byron P. (Barney) Everitt was presumably enjoying his retirement from the body manufacturing business (see Volume 2), he still had the urge to become involved again in building a fine car. So in 1920 Everitt met with Rickenbacker several times to make plans to build and introduce a new car to be called the "Rickenbacker." It was to be an innovative six-cylinder car, selling in the price range of $1,000 to $1,500.

William E. Metzger, Walter E. Flanders, and E. LeRoy Pelletier were invited to join their business venture. Flanders enthusiastically agreed, but Metzger declined as he was totally and successfully involved as a dealer of several makes of cars, and was about to become the Wills-Sainte-Claire distributor for the state of Michigan. Pelletier by this time had his own successful advertising agency, but he willingly joined Everitt, Flanders, and Rickenbacker as manager of advertising and sales promotion.

The "Rickenbacker"

The Rickenbacker Motor Company was organized and incorporated in the state of Michigan on July 25, 1921. The officers were Byron P. Everitt president, Edward V. Rickenbacker vice president of public relations, Walter E. Flanders vice president in charge of production, Harry Cunningham (former E-M-F dealer) secretary and treasurer, E.R. Evans chief engineer, and Cliff Durant director of sales. Late in 1921 about $5,000,000 of stock was issued and sold to about 13,000 stockholders. Everitt, Flanders, Rickenbacker, and Cunningham retained about 25% of the total.

Due to the depression of 1920-21, there was plenty of vacant factory space available. The Rickenbacker Motor Company purchased a large plant at 4815 Cabot Avenue near Michigan Avenue in Detroit from the Disteel Wheel Company for a very low price. The engineering and design progressed favorably during 1921, under the direction of E.R. Evans. The Rickenbacker Six engine featured a seven-main-bearing counterbalanced crankshaft, twin flywheels (one at each end of the crankshaft), and vertical ignition distributor (above the cylinder head). The frame had a double-drop feature which provided a very low center of gravity for the low-slung body, aiding in the superlative handling characteristics of the car. The low-slung body design was very attractive and appealed to young car buyers. The prototypes were built and road-tested under all road conditions.

In the meantime during 1920 and 1921, to get first-hand selling experience, Rickenbacker established a distributorship for General Motors' newly introduced Sheridan car for the state of California. The experience was invaluable for Rickenbacker because he had to establish a market for an unknown product against the sales resistance caused by the depression of 1920-21. He corrected the conditions that caused the lack of sales, and successfully established 50 Sheridan dealers in the state.

Trouble from the Start

The new Rickenbacker Six was introduced at the National Auto Show in New York on January 7, 1922. The meticulously finished display models consisted of a touring car at $1,485, a coupe at $1,885, and a sedan at $1,995. Unfortunately, the three cars not only represented the entire Rickenbacker line but also the total production of the plant up to that date. Production started in March 1922 at ten cars per day. The high-performance engine and a low-slung body, among other selling features, made sense to the buying public, and

1925 Rickenbacker 6 brougham. (Source: David Rickenbacker)

the car began to sell easily. Eddie Rickenbacker criss-crossed the United States by plane several times, selling to dealers. At its peak, Rickenbacker had established about 1,200 dealers in the United States, and about 300 in foreign countries. Rickenbacker Motor Company could have enjoyed phenomenal growth; however, due to the delay in production, only about 3,709 cars were sold and delivered in 1922.

In July 1923 Rickenbacker introduced mechanical four-wheel brakes. Even though Duesenberg featured hydraulic four-wheel brakes two years before, and Packard introduced mechanical four-wheel brakes only a few weeks earlier, it was Rickenbacker that was the target of any competitors who didn't have four-wheel brakes to offer and had thousands of obsolete cars to sell. One major manufacturer went so far as to run full-page ads in newspapers across the nation attacking four-wheel brake systems as extremely dangerous. Competitive dealers and salesmen verbally ganged-up on Rickenbacker and its four-wheel brakes. Ironically, the major manufacturer who led the attack adopted four-wheel brakes five years later.

These unfair business tactics and adverse publicity affected Rickenbacker sales, although there were some intelligent buyers with enough technical knowledge to recognize the superiority of four-wheel brakes. Despite the unfair tactics, Rickenbacker sold 5,958 new cars in 1923. For 1924 Rickenbacker introduced the new Vertical "8" Superfine, featuring an eight-in-line engine with nine-main-bearing crankshaft and twin flywheel (a feature rediscovered in an Indy race car some 60 years later).

It was not only the slanderous attacks by its competitors that led to the floundering of Rickenbacker Motor Company, but also the sudden death of Walter E. Flanders on June 16, 1923, from an automobile accident near Newport News, Virginia. Even though the economy of the nation was improving, Rickenbacker was losing its dealers, either through bankruptcy or by the dealers switching to other makes of cars.

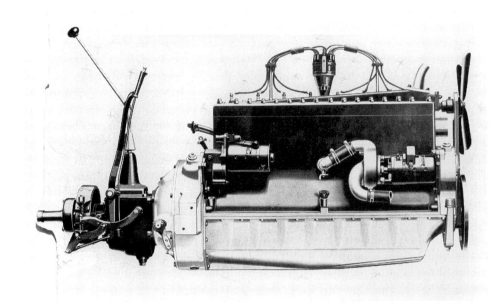

The 1925 Vertical "8" Superfine. (Source: Indianapolis Motor Speedway Corp.)

1925 Indianapolis 500 with Rickenbacker Vertical "8" as pace car. Eddie Rickenbacker driving, Fred Duesenberg passenger seat, Chester Ricker right rear, and T.E. Meyers left rear. (Source: Indianapolis Motor Speedway Corp.)

155

*1927 Rickenbacker Victoria.
(Source: Indianapolis Motor
Speedway Corp.)*

About 6,583 Rickenbacker cars were sold domestically in 1924, and the peak year of 1925 increased the sales to 8,049 units. For 1926 Rickenbacker introduced laminated safety windshield glass, whereby two sheets of plate glass were bonded together with a sheet of celluloid between them. Stutz at this time offered safety glass with steel wires molded in the glass horizontally, spaced about 2 in. apart.

During 1926 the management of Rickenbacker Motor Company started to squabble among themselves. Rickenbacker, in an effort to restore unity and accord among them, decided to leave the company. He wrote a brief note of resignation and departed; however, the plan backfired. The loss of Flanders and Rickenbacker was so profound that costs could not be controlled, product quality suffered, and sales tumbled to only 4,050 Rickenbacker cars in 1926.

For 1927 Rickenbacker cars were favorably redesigned, but it was too late. About 500 cars were built and distributed to Rickenbacker dealers and management officials, but the company went into receivership and bankruptcy. Byron Everitt, Harry Cunningham, Cliff Durant, and Edward Rickenbacker, along with all the individual stockholders of Rickenbacker Motor Company, lost their investments. Eddie Rickenbacker, in addition to the losses sustained in the bankruptcy, found himself unemployed at age 35, and a quarter million dollars in debt.

Starting Over

Rickenbacker may have lost his fortune, but not his popularity nor his splendid reputation of honesty and integrity. In 1926 Donald Douglas (former chief engineer of Glenn L. Martin Company) formed his own aircraft company in Long Beach, California. To get Donald Douglas started, Rickenbacker arranged with Harry Chandler (publisher of the Los Angeles Times) and his associates to subscribe for the capital stock to provide the operating capital.

Charles W. Schwab and other investment bankers had the confidence in Rickenbacker to loan him enough money to purchase the Indianapolis Motor Speedway from James Allison on August 17, 1927. Rickenbacker also arranged for the purchase of the Allison Engineering Company by the Fisher Brothers Investment Trust.

Left to right: Allison,
Rickenbacker, Schwab.
(Source: Indianapolis Motor
Speedway Corp.)

The Airline Industry

On January 1, 1928, Rickenbacker joined the Cadillac LaSalle Division of General Motors Corporation, and on July 1, 1929, became vice president of sales for GM's Fokker Aircraft Division. He was named vice president of American Airways on April 29, 1932, and was elected vice president of North American Aviation on February 28, 1933.

In an attempt to forestall a tragic mistake by the United States Post Office Department, Rickenbacker, on February 18, 1934, established a transcontinental flight record from Los Angeles to Newark, New Jersey, in 13 hours, 2 minutes. But the next day the U.S. Post Office Department cancelled the commercial airmail contracts and tried to use U.S. Army pilots to fly mail. Tragically, many of the pilots needlessly lost their lives. To save face the U.S. Post Office Department finally awarded airmail contracts to certain airlines with many restrictions.

In December 1934 Rickenbacker was requested to operate Eastern Airways division of North American Aviation. On January 1, 1935, he was elected general manager of Eastern Airlines (a change of name). Under his management Eastern Airlines prospered financially and grew with new terminals, routes, and equipment. In 1937 Eastern Airlines was awarded a safety award for no fatalities during 1930-36, with 141,794,894 passenger miles. In 1940 Eastern posted a profit of $1,575,456.

In the 27th race for his life (going back to his war years), Rickenbacker survived an airline plane crash near Atlanta on February 26, 1941, with some permanent injuries. In September 1942 he was offered a commission as Major General, but instead accepted a $1 appointment as civilian advisor to the U.S. armed forces. For an inspection tour he left Hickam Field (Pearl Harbor) on October 19, 1942, in a B-17-D bound for

Canton China. His plane was downed in the southwest Pacific on October 21, 1942. He was adrift in the Pacific for 24 days, and was rescued on November 14, 1942. That was his 28th and greatest race of his life. After recovery from the ordeal he served on the staff of Henry L. Stimson, Secretary of War, lecturing across the entire United States, spearheading the U.S. War Bond drive during the balance of World War II.

Rickenbacker sold the Indianapolis Motor Speedway to Anton Hulman in late 1945 for $700,000, exactly the price he paid for it in 1927. On October 1, 1959, he retired as president of Eastern Airlines, but remained on the board of directors as chairman until December 31, 1963. With all due respect to the subsequent management of Eastern Airlines, in retrospect it might still be in business today if it were managed by someone of the caliber of Edward V. Rickenbacker. Rickenbacker died on July 23, 1973, at the age of 82, while vacationing in Switzerland.

Left to right: Seth Klein, Eddie Rickenbacker, Eddie Edenburn, Theodore "Pop" Meyers. (Source: Indianapolis Motor Speedway Corp.)

E.L. Cord

E.L. Cord was one of the key players in the United States' development of the finest transportation systems in the world, encompassing both automobiles and aircraft. His associates Frederick S. Duesenberg, Lucius B. Manning, Roy H. Faulkner, Alan H. Leamy, and Gordon Buehrig were also responsible for his success.

Why then didn't the Cord Corporation become the fourth of an automotive Big "Four"? Was it because it arrived on the automotive scene too late, at the worst possible time, or in the wrong era? We can only ponder what we may have reaped if the Cord Corporation could have survived the merciless Depression.

* * *

The Origins of the Corporation

Eckhart Carriage Company

The Cord Corporation can trace its roots back to a man named Charles Eckhart, who was born in 1841 in Germantown, Pennsylvania. In 1857 Charles began as an apprentice in a carriage factory in Hilltown, Pennsylvania, where he soon learned most of the aspects of the wagon and carriage business. He saved his earnings, and with additional borrowed cash, he was able to purchase his employer's interest in the carriage

Charles Eckhart, founder of the Eckhart Carriage Company.

Morris Eckhart, co-founder, with his brother Frank, of the Auburn Automobile Company in 1900.

factory in 1860. His carriage manufacturing career was interrupted by his enlistment in the Union Army during 1861. Upon his release to inactive duty in 1865, Charles attempted to resume his education, but for financial reasons he returned to carriage building at a factory in Norristown, Pennsylvania.

During 1866, he visited northwest Indiana, where he met and married his wife Barbara. However, feeling there was more opportunity in eastern Pennsylvania, they settled in Chester County where Charles resumed his carriage-building trade. Their first son Frank was born in 1867. Anna, Morris, and William were born during the next seven years. Business was favorable until the national financial crisis in 1874, when Charles decided to move back to Indiana; he established residence in a two-story brick house that had an adjoining five acres of vacant land. After the purchase was arranged he planned to begin building carriages again, for the time being, processing the carriage components within the house.

Charles founded the Eckhart Carriage Company in 1874. As the sales of the buggies and surries increased, he expanded by erecting a two-story carriage factory in 1883 on the vacant property. In 1885 the Eckhart Carriage Company was incorporated in the state of Indiana. The manufacture of carriages, surries, and buggies was done on a production basis, and the distribution was by selling wholesale to implement, carriage, and buggy dealers.

In 1893, Charles Eckhart retired, but remained on the board of directors. His son Morris was elected president, and Frank was elected vice president. Morris assumed the management of the business, and Frank spent most of his time traveling to the major cities in the Midwest, selling wholesale to buggy and carriage dealers

The Auburn Automobile Company

Even though Morris and Frank foresaw the automobile replacing the horse-drawn vehicles, sales of the carriages, buggies, and surries was so successful that they were very hesitant about getting into the manufacture of automobiles. However, in 1900 they founded the Auburn Automobile Company with the blessing and financial support of their father. During 1900 Morris and Frank were experimenting with gasoline-engine-powered vehicles, and built their first prototype vehicle. It was a single-seat

The first product of the Auburn Automobile Company, the prototype runabout, c. 1902.

runabout powered by a single-cylinder gasoline engine under the seat. The drive was by chain to the rear axle, and the steering was by a tiller. The experimentation continued through 1902, as they built about two dozen cars, but by 1903, Morris and Frank knew they had to get into automobile manufacturing with full intent and all-out effort. While the experimental prototypes, of which there were about 20 to 30, may have been sold, production and sale of Auburn started in 1903.

While sales record were not kept, it is possible that a total of up to 100 cars may have been built and sold by the end of 1903. The Model A Auburn was continued through 1904, built in the new 60 by 100 ft. two-story plant, erected near the carriage factory building. Approximately 50 Auburn cars were built and sold during 1904.

Various new models were introduced through the years 1905-1911. See the tables for details.

	1903-1905 Model A	1903-1905 Model B
Wheelbase (in.)	78	96
Price	$1,000	$1,250
No. of Cylinders / Engine	1	2
Bore x Stroke (in.)	5.00 x 5.00	5.00 x 5.00
Horsepower	10 adv, 10 ALAM	18 adv, 20 ALAM
Body Styles	runabout, tonneau, touring	touring
Other Features		

	1906 Model C	1907 Model D
Wheelbase (in.)	94	100
Price	$1,250	$1,250
No. of Cylinders / Engine	2	2
Bore x Stroke (in.)	5.00 x 5.00	5.00 x 5.00
Horsepower	24 adv, 20 ALAM	24 adv, 20 ALAM
Body Styles	touring	touring
Other Features		

1904 Auburn Model A rear-entrance tonneau, the earliest Auburn extant. Restored by a grant from the Institute of Museum Services. (On display at the Auburn Cord Duesenberg Museum.)

	1908 Models G, H, K	**1909 Models G, H, K**
Wheelbase (in.)	100	100
Price	$1,250-1,350	$1,250-1,350
No. of Cylinders / Engine	2	2
Bore x Stroke (in.)	5.00 x 5.00	5.00 x 5.00
Horsepower	24 adv, 20 ALAM	24 adv, 20 ALAM
Body Styles	runabout, touring	runabout, touring
Other Features		

	1909 Models B, C, D	**1910 Models G, H, K**
Wheelbase (in.)	106	100
Price	$1,350-1,400	$1,250-1,350
No. of Cylinders / Engine	4	2
Bore x Stroke (in.)	4.25 x 4.75	5.00 x 5.00
Horsepower	30 adv, 28.9 ALAM	24 adv, 20 ALAM
Body Styles	runabout, touring	runabout, roadster, touring
Other Features		

	1910 Models B, C, D	**1910 Models R, S, X**
Wheelbase (in.)	106	116
Price	$1,350-1,400	$1,650
No. of Cylinders / Engine	4	4
Bore x Stroke (in.)	4.25 x 4.75	4.50 x 5.00
Horsepower	30 adv, 28.9 ALAM	40 adv, 32.4 ALAM
Body Styles	runabout, roadster, touring	runabout, roadster, touring
Other Features		

1910 Auburn Model X. This illustration was used in the 1910 Auburn sales catalog, "House Party at Coldwater Lake, Michigan."

	1911 Models G, K	**1911 Models F, L**
Wheelbase (in.)	100	106
Price	$1,250	$1,400
No. of Cylinders / Engine	2	4
Bore x Stroke (in.)	5.00 x 5.00	4.25 x 4.75
Horsepower	24 adv, 20 ALAM	30 adv, 28.9 ALAM
Body Styles	runabout, roadster, touring	runabout, roadster, touring
Other Features		

	1911 Models M, N	**1911 Models T, Y**
Wheelbase (in.)	116	116
Price	$1,650	$1,700
No. of Cylinders / Engine	4	4
Bore x Stroke (in.)	4.50 x 5.00	4.50 x 5.00
Horsepower	40 adv, 32.4 ALAM	40 adv, 32.4 ALAM
Body Styles	runabout, roadster, touring	runabout, roadster, touring
Other Features		

During this period Morris and Frank Eckhart continued to divide their interest between the Eckhart Carriage Company and the Auburn Automobile Company since they were not sure if the automobile would be just another fad (like the bicycle). At this point the carriage business was profitable, but the automobile was not.

	1912 Models 30L, 35L	**1912 Models 40H, 40N, 40M**
Wheelbase (in.)	116	120
Price	$1,100-1,400	$1,750
No. of Cylinders / Engine	4	4
Bore x Stroke (in.)	4.25 x 4.75	4.50 x 5.00
Horsepower	30 adv, 28.9 NACC	40 adv, 32.4 NACC
Body Styles	runabout, roadster, touring	runabout, roadster, touring
Other Features		

1913 Auburn Model 6-45 touring car.

	1913 Model 33L	1913 Model 37L
Wheelbase (in.)	116	116
Price	$1,250	$1,400
No. of Cylinders / Engine	4	4
Bore x Stroke (in.)	3.75 x 5.25	4.25 x 4.75
Horsepower	30 adv, 22.5 NACC	33 adv, 28.9 NACC
Body Styles	touring, roadster, coupe	touring, roadster, coupe
Other Features		

	1913 Model 40L	1913 Model 6-45
Wheelbase (in.)	116	126
Price	$1,650	$2,150
No. of Cylinders / Engine	4	6
Bore x Stroke (in.)	4.50 x 5.00	3.75 x 5.25
Horsepower	40 adv, 32.4 NACC	45 adv, 33.7 NACC
Body Styles	touring, roadster, coupe	touring, roadster, coupe
Other Features		

	1913 Model 6-50	1914 Model 4-40
Wheelbase (in.)	135	120
Price	$3,000	$1,490
No. of Cylinders / Engine	6	4
Bore x Stroke (in.)	4.12 x 5.25	4.50 x 5.00
Horsepower	50 adv, 40.8 NACC	40 adv, 32.4 NACC
Body Styles	touring, roadster, coupe	touring, roadster, coupe
Other Features		

	1914 Model 4-41	**1914 Model 6-45**
Wheelbase (in.)	120	126
Price	$1,540	$2,000-2,100
No. of Cylinders / Engine	4	6
Bore x Stroke (in.)	4.50 x 5.00	3.75 x 5.25
Horsepower	40 adv, 32.4 NACC	45 adv, 33.7 NACC
Body Styles	touring	touring, roadster
Other Features		

	1914 Model 6-46	**1915 Model 4-36**
Wheelbase (in.)	126	114
Price	$2,100	$1,075
No. of Cylinders / Engine	6	4
Bore x Stroke (in.)	3.75 x 5.25	3.75 x 5.00
Horsepower	45 adv, 33.7 NACC	36 adv, 22.5 NACC
Body Styles	six-passenger touring	touring, roadster, coupe
Other Features		

	1915 Model 6-40	**1915 Model 6-47**
Wheelbase (in.)	120	127
Price	$1,550	$2,000
No. of Cylinders / Engine	6	6
Bore x Stroke (in.)	3.50 x 5.00	3.50 x 5.25
Horsepower	40 adv, 29.4 NACC	47 adv, 29.4 NACC
Body Styles	touring, roadster, coupe	touring, roadster, coupe
Other Features		

In 1915 Charles Eckhart passed away following a lengthy illness at the age of 74. Because Morris and Frank Eckhart had been directing the management of Auburn Automobile Company for several years, there were no changes of company policy.

	1916 Model 4-38	**1916 Model 6-38**
Wheelbase (in.)	114	120
Price	$985	$1,050
No. of Cylinders / Engine	4	6
Bore x Stroke (in.)	3.87 x 5.00	3.00 x 5.00
Horsepower	38 adv, 24.0 NACC	38 adv, 21.6 NACC
Body Styles	touring, roadster	touring, roadster
Other Features		

	1916 Model 6-40A	**1917 Model 6-39**
Wheelbase (in.)	127	120
Price	$1,375	$1,145
No. of Cylinders / Engine	6	6
Bore x Stroke (in.)	3.50 x 5.25	3.00 x 5.00
Horsepower	40 adv, 29.4 NACC	39 adv, 21.6 NACC
Body Styles	touring	touring, roadster
Other Features		

	1917 Model 6-44	**1918 Model 6-39**
Wheelbase (in.)	131	120
Price	$1,535	$1,145-1,535
No. of Cylinders / Engine	6	6
Bore x Stroke (in.)	3.50 x 5.25	3.12 x 4.50
Horsepower	44 adv, 29.4 NACC	39 adv, 23.4 NACC
Body Styles	touring, roadster	touring, roadster
Other Features		

	1918 Model 6-44	**1919 Model 6-39H**
Wheelbase (in.)	131	120
Price	$1,685	$1,695-2,475
No. of Cylinders / Engine	6	6
Bore x Stroke (in.)	3.50 x 5.25	3.25 x 4.50
Horsepower	44 adv, 29.4 NACC	40 adv, 25.3 NACC
Body Styles	touring, roadster	touring, roadster, coupe, sedan
Other Features		

Changing Hands

By 1918, the Eckharts realized that the automobile was here to stay, and they dissolved The Eckhart Carriage Company.

During World War I, restrictions adversely affected Auburn production and sales. Distraught by the nagging problems of high inflationary costs, sagging sales, and the downward spiral of the nation's economy, Morris and Frank Eckhart decided that they would sell the Auburn Automobile Company and retire if a buyer could be found. It so happened the Eckharts found such a buyer: a financial consortium headed by Ralph A. Bard of Hitchcock and Company, executives of two banking firms in Chicago, and William K. Wrigley Jr. were interested. In early 1919 the group made a deal with the Eckharts, but with the stipulation that Morris Eckhart would remain and manage the Auburn Automobile Company until a successor could be found. The new owners' first move was to call the car the Auburn "Beauty Six."

	1920 6-39H	**1921 6-39K**	**1922 Beauty Six**
Wheelbase (in.)	120	120	121
Price	$1,695-2,475	$1,895-2,995	$1,575-2,395
No. of Cylinders / Engine	6	6	6
Bore x Stroke (in.)	3.25 x 4.50	3.25 x 4.50	3.25 x 4.50
Horsepower	43 adv, 25.3 NACC	43 adv, 25.3 NACC	55 adv, 25.3 NACC
Body Styles	4 plus sport touring	4	4
Other Features			

On June 21, 1922, A.P. Kemp was elected president and general manager of the Auburn Automobile Company to succeed Morris Eckhart effective July 1, 1922. Kemp had been vice president and treasurer. Other officers elected were: J.I. Farley vice president, E.A. Johnson secretary, Z.B. Walling assistant secretary, and John Zimmerman assistant treasurer. It was understood that Morris Eckhart had planned to travel extensively in Europe and the United States during the next two years.

Just one year later, J.I. Farley was elected president of Auburn Automobile Company, and in December 1923, Roy Faulkner was advanced from sales manager to director of sales. E.H. Gilcrest was named sales manager succeeding Faulkner. During 1923, Auburn domestic sales amounted to approximately 2,443 cars.

1923 Auburn Automobile Company factory. Final assembly of the Beauty Six line of touring cars and sedans. This photo was featured in the company's 1923 franchise book.

	1923 Model 6-51	**1924 Model 6-43**	**1924 Model 6-63**
Wheelbase (in.)	121	114	124
Price	$1,475-1,995	$1,095-1,595	$1,595-2,345
No. of Cylinders / Engine	6	6	6
Bore x Stroke (in.)	3.37 x 4.50	3.12 x 4.25	3.25 x 5.00
Horsepower	55 adv, 27.3 NACC	43 adv, 23.4 NACC	60 adv, 25.3 NACC
Body Styles	4 plus sport touring	4	4
Other Features			

E.L. Cord Takes Auburn to New Heights

On September 6, 1924, Errett Lobban Cord, former vice president and sales manager of Quinlan Motors Company of Chicago, joined the Auburn Automobile Company. (It was ironic that Stewart McDonald and Frederick Rengers of Moon Motor Car Company did not recognize the "brilliant diamond" they had in E.L. Cord at Quinlan. Moon acquired the assets of Quinlan Motors Company in 1926. It was like buying the stable after the champion race horse was gone.) Just two-and-a-half weeks later Cord was elected vice president and general manager of Auburn Automobile Company. According to production records, there were approximately 3,954 Auburn cars built, but new car registrations showed only 2,474 cars sold during

Auburn Automobile Company body fitting department, c. 1923.

1924 Auburn 6-63 sedan.

1924. Hence, there were hundreds of unsold new cars parked in the factory and dealer's lots. Cord's first move was to paint the unsold cars in attractive bright colors, and have the radiator shell, headlamps, and other exterior parts nickel-plated. The cars quickly sold, but not at a profit; instead, the deficit for 1924 was about $69,000.

Cord's next move was to lengthen the wheelbase of the 6-63 to 129 in., and to equip the chassis with a new Lycoming straight-eight engine (IL-8). This model was called the Auburn 8-63, and was publicly announced on September 21, 1924.

	1925 Model 6-43	1925 Model 8-63
Wheelbase (in.)	114	129
Price	$1,395-1,945	$1,895-2,550
No. of Cylinders / Engine	6	IL-8
Bore x Stroke (in.)	3.12 x 4.25	3.12 x 4.25
Horsepower	50 adv, 23.4 SAE	65 adv, 31.2 SAE
Body Styles	5	touring, brougham, sedan
Other Features		balloon tires, four-wheel brakes

	1926 Model 4-44	1926 Model 6-66
Wheelbase (in.)	120	120
Price	$1,145-1,195	$1,395-1,795
No. of Cylinders / Engine	4	6
Bore x Stroke (in.)	3.62 x 5.00	3.12 x 4.50
Horsepower	42 adv, 21.0 SAE	50 adv, 25.3 SAE
Body Styles	touring, roadster, coupe, sedan	roadster, brougham, sedan
Other Features	balloon tires, hydraulic four-wheel brakes	

	1926 Model 8-88	1927 Model 6-66A
Wheelbase (in.)	129	120
Price	$1,695-2,095	$1,095-1,345
No. of Cylinders / Engine	IL-8	6
Bore x Stroke (in.)	3.25 x 4.50	2.87 x 4.75
Horsepower	68 adv, 33.8 SAE	43 adv, 19.8 SAE
Body Styles	5 plus seven-passenger sedan	5
Other Features		

Enjoying domestic sales of about 4,044 cars in 1925, Auburn Automobile Company showed a profit of about $655,432.

For 1926, Auburn continued models 6-66 and 8-88; however, the styling was completely new. A new nickel-plated radiator shell, nickel-plated drum-shaped headlamps, newly shaped body panels, and the body belt molding curved upward over the hood upper panel to meet the opposite side belt molding at the radiator cap. This styling feature continued on all Auburns through the 1933 models. On December 24, 1925, the Auburn prices for the 6-66 and 8-88 models were drastically reduced, and at the same time, the new Auburn 4-44 was introduced, but discontinued after 1926.

Cord Becomes President

E.L. Cord was elected president of Auburn in early 1926. Sales that year were approximately 7,138 cars and the profit was about $686,559. Also, beginning in 1926 all future Auburn cars were to be powered by Lycoming engines.

The Acquisition of Duesenberg and Lycoming

On October 6, 1926, E.L. Cord and associates completed negotiations through Manning and Company (Lucius B. Manning, president) to take over the assets, control, and operation of Duesenberg Motors Company of Indianapolis, Indiana. The new Duesenberg organization was incorporated in the state of Indiana, capitalized at $1,000,000, and it became known as the Duesenberg Incorporated, a subsidiary of Auburn Automobile Company. The officers were E.L. Cord, president, and Frederick S. Duesenberg, vice president and chief engineer. During the interim period of 1927 and 1928, Duesenberg Incorporated built the Duesenberg Model X. (See the Duesenberg chapter in Volume 2.)

During 1927, Auburn domestic sales amounted to approximately 9,835 cars, generating a profit of approximately $912,704.

On September 6, 1927, E.L. Cord bought the Lycoming Manufacturing Company (and also the Limousine Body Company, since Auburn Automobile Company bought most of Limousine's production of bodies). Cord saw something more in the purchase of Lycoming than just passenger car engines. Since the successful flight of Charles A. Lindbergh to Paris in May 1927, there was a rekindled interest in aviation. Cord reasoned that if Lycoming could build reliable passenger car and truck engines, it certainly could build outstanding aircraft engines. While Lycoming also furnished engines for Elcar, Gardner, and Locomobile, the subsequent successful building of aircraft engines and the growth of the aircraft industry confirmed E.L. Cord's belief.

	1927 Model 8-77	**1927 Model DL8-88**
Wheelbase (in.)	125	130-146
Price	$1,395-1,745	$1,995-2,595
No. of Cylinders / Engine	IL-8	IL-8
Bore x Stroke (in.)	2.75 x 4.75	3.25 x 4.50
Horsepower	65 adv, 24.2 SAE	68 adv, 33.8 SAE
Body Styles	6	6
Other Features		

	1928 Model 6-76	**1928 Model 8-88**
Wheelbase (in.)	120	125
Price	$1,195-1,395	$1,495-1,695
No. of Cylinders / Engine	6	IL-8
Bore x Stroke (in.)	2.87 x 4.75	2.87 x 4.75
Horsepower	60 adv, 19.8 SAE	88 adv, 26.4 SAE
Body Styles	4	5 plus speedster
Other Features		

	1928 Model 115	**1929 Model 6-80**
Wheelbase (in.)	130	120
Price	$1,995-2,395	$995-1,095
No. of Cylinders / Engine	IL-8	6
Bore x Stroke (in.)	3.25 x 4.25	2.87 x 4.75
Horsepower	115 adv, 33.8 SAE	70 adv, 19.8 SAE
Body Styles	5 plus speedster	6
Other Features		

1928 Auburn 8-88 Speedster.

	1929 Model 8-90	1929 Model 115	1929 Duesenberg J
Wheelbase (in.)	125	130	134-141
Price	$1,395-1,695	$1,995-2,395	$8,500
No. of Cylinders / Engine	IL-8	IL-8	IL-8
Bore x Stroke (in.)	2.87 x 4.75	3.25 x 4.50	3.75 x 4.75
Horsepower	100 adv, 26.4 SAE	125 adv, 33.8 SAE	265
Body Styles	7	8	custom
Other Features			

In January 1928, the new Auburn "115" made its debut at the National Auto Show in New York, developing 115 hp. Most of the increase in power was the result of new manifolding and bypass muffler, dual carburetion, and increased compression ratio. But the most exciting cars were the Auburn 8-88 and Auburn 115 speedsters, which also made their debut.

On April 23, 1928, Auburn got into production of the cabriolet in the 76, 88, and 115 models, which were a combination of open and closed body styles. Record sales of the 88 and 115 required a cutback of the 76 model. Sales continued briskly through May and June, as Auburn shipped 1,500 cars during June. Auburn domestic sales amounted to approximately 11,157 cars during 1928.

The Lycoming Manufacturing Company experienced record sales and production of Lycoming engines, so much so that many departments had to operate on a 24-hour basis, employing more than 2,000 workers. In addition to the domestic customers of Lycoming engines, Crossley Ltd. of England placed an order on July 5 for 2,000 engines (they were previously supplied by Vauxhall of England).

On June 30 and July 1, 1928, two Auburn 115 speedsters established twelve new stock car records at the Atlantic City Motor Speedway. The record results were 85.49 mph for 5 miles, to 84.69 mph for 2,000 miles. Also, the records included 1 hour at 85.59 mph and 24 hours at 84.73 mph. The Auburn speeds exceeded the former speed records by 5-11 mph.

Meanwhile, Fred S. Duesenberg was busy designing the magnificent Duesenberg J, and Cornelius Van Ranst, Harry Miller, and Leo Goosen were hard at work to translate the Miller front-drive principles to a viable

1929 Auburn cabin speedster.

passenger car. Herbert Snow, Auburn's chief engineer, and John Oswald, body designer, were also involved in these projects, as was Alan Leamy, a brilliant young designer who became the chief stylist for the front-wheel-drive project. Prior to his joining Auburn, Leamy was the stylist for the Marmon Big Eight during 1927 and 1928. The aim was to have these two outstanding cars ready for the 1929 National Automobile Show in New York in January. The New Duesenberg J was without question the hit of the show, earning glowing compliments, honors, and accolades. The front-drive project car would wait until June to make its debut.

The Formation of Cord Corporation

On January 15, 1929, the Auburn Automobile Company purchased the Central Manufacturing Company. Ellis Ryan, former vice president of Central Manufacturing Company, was elected vice president of the Auburn Automobile Company, in charge of the operations in Connersville, Indiana. E.L. Cord purchased the buildings and real estate of the following: former Ansted Engine Company, Lexington Motor Company, McFarlan Motors Company, and many other idle plants in Connersville, Indiana. The city officials of Connersville apparently made E.L. Cord an offer "he couldn't refuse."

The reported purchase included 20 buildings with a combined floor space of approximately 1,500,000 sq. ft., and real estate of about 82 acres, at a reported purchase price of $2,000,000.

Also in 1929 Cord purchased a 140-acre farm north of Connersville, Indiana, with the intent of using it as an aeroplane factory and landing field. Always remaining flexible if needed, instead of building an aeroplane factory, he bought the Stinson Aircraft Corporation of Wayne, Michigan. Stinson had been an engine customer of Lycoming.

On June 21, 1929, E.L. Cord announced the formation of the Cord Corporation. Incorporated with an indicated capitalization of $125,000,000, it became the holding company for the subsidiaries and other E.L. Cord holdings. The officers elected were: E.L. Cord president, L.B. Manning vice president, R.S. Pruitt secretary, and Hayden Hodges treasurer. With eight divisions becoming the nucleus, 52 subsidiary companies were also under the umbrella of the Cord Corporation by the end of 1929.

The Cord Front-Drive Car

The long-awaited front-drive passenger car was announced by the Auburn Automobile Company on June 29, 1929. Named the "Cord" after the company president, it represented one of the most outstanding developments of passenger car design in many years, not only because it was a front-drive car but also because in it were combined departures from accepted practice, which set the car apart as "not just another car." This fact was best emphasized by the body, while conforming to modern accepted standards of beauty, it evinced just enough of the unusual to make its appearance one of distinct individuality, due to the skill and talent of Alan Leamy.

1929 Cord L29 sedan assembly line at the Auburn Automobile Company, Auburn, Indiana. Car bodies are dropped from above.

Designed around the Harry Miller front-drive patents, Cord took the lead in making available to the public the front-drive principle. Many innovative features were evident in the new Cord L29, which had been under experimentation and tests for over two years. The sharp-angled V-shaped radiator shell and shutters was something new to American cars. The front fenders were longer and more sweeping than those found on other production automobiles. They were in keeping with the exceptional long and tapering hood to house the driving units of the car. The bodies extended out and down to the running board, with practically no splash guard ledge. While the bodies were approximately 10 in. lower than the conventional rear-drive cars, there was ample head room and the doors were sufficiently high to permit reasonable ease of entry and egress.

The entire design reduced the unsprung weight, which was reflected in the better riding qualities. The elimination of the rear axle kick-up and the unconventional frame bracing enabled the use of a light body, resulting in a total lighter weight of only about 3,975 lb.

Being a front-wheel drive, all power transmission units were located ahead of the dash panel, with the general order reversed. The engine, clutch, transmission, and differential were all bolted together in a unit power plant. In operation the power was transmitted to the driving front wheels by means of two driveshafts, each having two universal joints. The wheels themselves were pivoted to a tubular dead axle, which was bowed forward in an arc to clear the driving shafts and differential housing. Both drive and torque reaction were taken through the double quarter-elliptic front springs. The springs were mounted with one spring above and one below the frame side channel. The forward ends of the springs were attached to the axle by means of a bushed yoke.

After the introduction of the Cord L29, due to the favorable publicity by the news media and automotive trade journals, the natural interest in the automobile industry, and the excellent promotion by E.L. Cord, the Auburn dealers' salesrooms were besieged by swarms of automobile shoppers and buyers. At least 2,000 orders for the new Cord were taken by the dealers, some of the orders were by buyers who had not even seen it yet.

Black Thursday and the Beginning of the End

Unfortunately, Auburn Automobile Company was faced with the difficulty of getting the Cord production going soon enough. Too soon "Black Thursday," October 24, 1929, descended on the American economy; thus, many prospects did not become buyers. As a result, the 1929 domestic sales of the Cord L29 amounted to only about 799 units. The Cord L29 models continued through 1930 and 1931, but at reduced prices.

Meanwhile, other divisions of the Cord Corporation were operating profitably, namely the Lycoming Manufacturing Company and Duesenberg Incorporated. In spite of the gloom caused by Black Thursday and the subsequent Depression, there was enough momentum in Duesenberg sales that Duesenberg Incorporated showed a profit of $84,055 for the year of 1929. Kings, Monarchs, royalty family members, as well as many famous American movie stars and successful businessmen preferred to drive a Duesenberg over the finest and most expensive cars the European manufacturers could offer.

While all stocks on Wall Street tumbled immediately following Black Thursday, six days later on October 30, 1929, the Auburn Automobile Company stock rebounded from 130 to 180. In spite of the worsening business conditions and the increased spending in expansion investment, Auburn Automobile Company enjoyed domestic sales of approximately 18,652 cars in 1929. With the help of the other profitable divisions, the Cord Corporation showed a profit of approximately $3,603,200 for 1929.

1930-1931

	1930 Model 685	1930 Model 895
Wheelbase (in.)	120	125
Price	$995-1,095	$1,195-1,395
No. of Cylinders / Engine	6	IL-8
Bore x Stroke (in.)	2.87 x 4.75	2.87 x 4.75
Horsepower	70 adv, 19.8 SAE	100 adv, 26.4 SAE
Body Styles	3	4
Other Features		

	1930 Model 125	1930 Cord L29
Wheelbase (in.)	130	137.5
Price	$1,445-1,695	$3,095-3,295
No. of Cylinders / Engine	IL-8	IL-8
Bore x Stroke (in.)	3.25 x 4.50	3.25 x 4.50
Horsepower	125 adv, 33.8 SAE	125 adv, 33.8 SAE
Body Styles	4	sedan, brougham, cabriolet, phaeton
Other Features		

	1931 Model 8-98	1931 Model 8-98A	1931 Cord L29
Wheelbase (in.)	127-136	127-136	137.5
Price	$945	$1,195	$2,395
No. of Cylinders / Engine	IL-8	IL-8	IL-8
Bore x Stroke (in.)	3.00 x 4.75	3.00 x 4.75	3.25 x 4.50
Horsepower	98 adv, 28.8 SAE	98 adv, 28.8 SAE	125 adv, 33.8 SAE
Body Styles	5	5	4
Other Features	Constant-mesh, silent-second transmission and a free-wheeling unit controlled by a short lever. Steeldraulic four-wheel brakes on 17-in.-diameter wood-spoke wheels, mounted with 6.00x17 balloon tires having a 6-in. cross-section (6.50 in. on the seven-passenger sedan).		

1930 Cord custom coupe.

Since the Auburn, Cord, and Duesenberg offerings were so far ahead of the competitors, no major technical changes were made for 1930; however, the model designations were changed to; 685, 895, and 125, superseding the models 6-76, B-88, and 115, respectively. A Cord cabriolet was chosen to pace the start of the 1930 Indianapolis 500 race. The pace car was driven by Wade Morton, who had driven many races in a Duesenberg racing car.

Sadly, during 1930, E.L.'s wife Helen passed away after many weeks of illness. Mrs. Cord's untimely passing had a profound effect on E.L. Cord and he literally buried himself in his work.

In spite of the Depression and business downturn during 1930, Auburn Automobile Company was able to sell approximately 13,149 cars domestically, of which about 1,416 were Cord L29s. For the fiscal year ending November 30, 1930, Auburn Automobile Company earned a profit of $1,018,331 and the Cord Corporation combined total profit was $1,477,477.

On December 3, 1930, the Lycoming Manufacturing Company announced the introduction of a new Lycoming V-12 marine engine. This new engine had an unusual 70-degree angle between the cylinder banks. The engine had a bore of 4-1/2-in. and a stroke of 4-1/2-in., giving a 1010-cu.-in. displacement, developing 300 hp. While this engine was most used for fast runabouts, it also powered commuter boats and express cruisers.

On January 3, 1931, the new 1931 Auburn 8-98 made its debut at the National Automobile Show in New York. It consisted of one line in a standard and series, having a distinctive styling and many mechanical innovations. It was totally designed and styled by Alan H. Leamy, and therefore resembled the Cord L29. The Auburn 8-98 had a broad, less-severe V radiator shell with vertical louvres instead of shutters, long sweeping front fenders, Cord style bodies, and sloping windshield pillars. The characteristic Auburn body belt molding continued diagonally across the upper hood panel, meeting the opposite side molding at the radiator shallow filler cap. The Auburn 8-98 was one of the first American cars to provide a channeled cross member in the center to brace the frame.

1931 Cord cabriolet.

During 1931, Auburn sales were so phenomenal that many long-established dealers of competitive cars abandoned their franchises and joined Auburn. The Auburn dealer count increased from approximately 650 dealers in 1930 to approximately 1,000 dealers by the end of 1931.

The Formation of Century Air Lines

On December 22, 1930, the aviation group of the Cord Corporation announced the formation of an enormous air transportation system to be known as the Century Air Lines Incorporated. Headquartered in Chicago, it was to have airline routes radiating from Chicago to Detroit, Toledo, Cleveland, and St. Louis.

March 23, 1931, marked the formal entrance of the Cord Corporation Century Air Lines into the transportation field. The occasion was heralded by impressive ceremonies in all six terminal cities. In Chicago, eight Stinson tri-motored airliners were christened by eight Chicago society women. On Thursday, March 26, and Friday March 27, similar ceremonies were held in the other five terminal cities. The Century Air Lines would provide frequent flights between the terminal cities, at airfare rates comparable to Pullman rail transportation. A commuter route was also established from Connersville, Indiana, to Chicago, Illinois, for the commuting of E.L. Cord and other Cord Corporation personnel.

Personnel Changes

At the urging of his friends and associates, E.L. Cord married Virginia Kirk Tharpe on January 3, 1931. They made home a penthouse in Chicago near the Cord Corporation headquarters. Later they moved to their estate near Los Angeles. Ten months after their wedding the Cords were blessed with a beautiful baby daughter. (Cord had two sons from his previous marriage.)

On January 19, 1931, W.H. Beal, former vice president and general manager, was elected president of Lycoming Manufacturing Company, succeeding John H. McCormick who resigned.

Roy H. Faulkner was elected president of Auburn Automobile Company on February 4, 1931, as E.L. Cord became chairman of the board of directors. Mr. Faulkner entered the automobile business as retail salesman for the Oakland Motor Car Company of Pittsburgh, Pennsylvania. He later became sales manager for the Reo distributor, then sales manager for the Stutz distributor, and finally general manager of the Frank Santry Motor Company of Cincinnati, Ohio, a Nash distributor. Certainly Mr. Faulkner's experience qualified him well to join Auburn Automobile Company in 1922 as sales manager, a position he held when E.L. Cord joined Auburn in 1924. Faulkner's elevation to the presidency of Auburn had been foreshadowed by the increasing responsibilities that Cord had been placing on him during the previous two years.

Roy Faulkner, sales manager, vice president, then president of the Auburn Automobile Company, 1922-1937. (Source: ACD)

However, in late November of that same year, Faulkner resigned from the presidency. Cord resumed as president and chief operating manager for the time being. L.B. Manning was named executive vice president. While Faulkner's plans for the future were not revealed, there had always been a mystery as to his departure because it seemed to have Cord's blessing. Auburn sales remained high during 1931, whereas the automobile industry as a whole suffered a severe loss of sales between 1929 and 1931. The industry total sales for 1931 amounted to 1,908,141 units, against 3,880,240 cars in 1929, a loss of approximately 50%. Auburn sales for 1931 amounted to approximately 30,952 cars (of which 335 were Cord sales), an increase of approximately 60% over the record year of 1929. The Cord Corporation enjoyed a profit of $3,579,849 and a surplus of $1,708,307 during 1931.

At the end of 1931, the divisions of the Cord Corporation included: Auburn Automobile Company; Duesenberg Incorporated; Stinson Aircraft Corporation; Spencer Heater Company; Central Manufacturing Company; Limousine Body Company; Columbia Axle Company; Century Airlines Incorporated, Chicago; Century Pacific Airlines Limited, Los Angeles; L.G.S. Devices Corporation; and approximately 50 subsidiary companies.

1932

	1932 Model 8-100	**1932 Model 8-100A**
Wheelbase (in.)	127	127
Price	$675-845	$805-975
No. of Cylinders / Engine	IL-8	IL-8
Bore x Stroke (in.)	3.00 x 4.75	3.00 x 4.75
Horsepower	100 adv, 28.8 SAE	100 adv, 28.8 SAE
Body Styles	coupe, phaeton, sedan, speedster	coupe, phaeton, sedan, speedster
Other Features		

	1932 Model 12-160	**1932 Model 12-160A**	**1932 Cord L29**
Wheelbase (in.)	133	133	137.5
Price	$975-1,145	$1,105-1,275	$2,395-2,595
No. of Cylinders / Engine	V-12	V-12	IL-8
Bore x Stroke (in.)	3.12 x 4.25	3.12 x 4.25	3.25 x 4.50
Horsepower	160 adv, 46.8 SAE	160 adv, 46.8 SAE	125 adv, 33.8 SAE
Body Styles	coupe, phaeton, sedan, speedster	coupe, phaeton, sedan, speedster	
Other Features			

The 1932 Auburn models were introduced on January 1, 1932, when E.L. Cord astounded the automotive industry by offering models with a new V-12 engine at prices starting at $975. The 1932 Auburn V-12 was also available as the Auburn Custom Twelve with dual ratio rear axle, priced at $1,105-1,275. But the star of the National Automobile Show in New York was the Auburn 12-160A speedster model priced at $1,275.

The Lycoming V-12-cylinder engine had horizontal valves in a lateral plane, actuated by a single camshaft operating against rocker levers, whose lash was taken up hydraulically. Lycoming spent several years developing this remarkable engine; Cornelius W. Van Ranst was the designer and chief engineer. L.G.S. freewheeling, Startix automatic starting, Bijur automatic chassis lubrication, and adjustable ride control were featured on the 1932 models.

During the week of July 4, 1932, two Auburn 12-160s, a speedster and sedan, broke 26 stock car records on Muroc Dry Lake, California. The distances ranged from 1 kilometer to 500 miles, in both the open and closed car classes. The fastest time was 95.1 mph by the Auburn 12-160 speedster and 91.08 mph for the closed car. The runs were electrically timed and officially recorded by the Contest Board of the American Automobile Association.

Tragically, Frederick S. Duesenberg died on July 26, 1932, as the result of an automobile accident in Pennsylvania.

On November 12, 1932, E.L. Cord, who had been serving as president, made the announcement of the election of W. Huber Beal to the presidency of Auburn Automobile Company. Beal had been president of Lycoming Manufacturing Company since January 1931, and continued in that position, thus dividing his time between Auburn, Indiana, and Williamsport, Pennsylvania, commuting on a Cord Corporation-owned Stinson airplane. On November 22, 1932, Harold T. Ames, president of Duesenberg Incorporated, became assistant to the Auburn president.

Despite the tremendous values of the Auburn Automobile Company's offerings and the technical innovations, the best Auburn could muster in 1932 was approximately 11,981 domestic car sales. This could be expected as the whole automobile industry agonized during the cruel Depression; the industry total sales amounted to only 1,096,399 cars in 1932, about half of the sales in 1931 and approximately one-quarter of the sales in 1929.

Control of Aviation Corporation

During the week of November 14, 1932, over the hum of engines speeding passengers across America's important air route, the rumblings of a fiscal "battle of the century" could be heard brewing, scheduled to reach a pitched encounter on December 11, when stockholders of the Aviation Corporation were scheduled to meet with the board of directors. E.L. Cord, chairman of the Cord Corporation and a director of Aviation Corporation, opened fire on the plans of the Aviation Corporation to "dilute its stock" to buy more properties at what he termed "less than half" of the proposed purchase price.

Cord was campaigning to increase the directorate of the corporation to 48. He had written the 28,000 stock-holders for their proxies to win control of the corporation and its 17 subsidiaries, by increasing the directorate and electing a majority to the board. The plan provided for the acquisition of the Aviation Corporation, assets of North American Aviation, Sperry Gyroscope, Eastern Air Transport, B/J Aircraft Corporation, and several subsidiaries, and the company's cash with the exception of $1,000.000.

The assets to be acquired included: a 25% interest in the Douglas Aircraft Company, a 10% interest in the Curtiss-Wright Corporation, and a 26% interest in the Transcontinental Air Transport, and a 1-1/2% increase of additional Pan-American Airways Corporation stock, increasing American Airways Corporation holdings in Pan-American Airways Corporation to 14%.

Cord, a 30% owner of the Aviation Corporation's outstanding stock and a board director, marshaled his efforts and his associates against LaMotte T. Cohu, president of Aviation Corporation, and a majority of its 35 directors. Lucius B. Manning, Cord Corporation executive vice president of financial operations, had been directing the contest for the proxies from New York while Cord had been supervising the action from Los Angeles. Cord obtained a temporary injunction to block immediate action on purchase.

The Aviation Corporation was incorporated in Delaware in 1929, the total assets on January 1, 1932, were $11,682,000, of which approximately $9 million was U.S. government bonds, cash, and other quick assets. Its subsidiaries operated five commercial airlines and ten contract airlines, including Century Air Lines. The American Airplane and Engine Corporation and Aviation Patent and Research Corporation were included in 17 subsidiaries, the consolidation of 81 companies.

By mid-March 1933, Cord advised the directors of Aviation Corporation that he and his associates had greatly increased their stock holdings to have effective control, that he would take full responsibility for the direction of the corporation's affairs, and that the corporation offices would be moved to Chicago. Richard F. Hoyt, president, and directors Cohu, Hann, Harriman, Lawrence, Morton, and Sloan resigned at that time. The campaign waged by E.L. Cord for control of Aviation Corporation apparently was brought to a successful conclusion with the election of L.B Manning to the presidency. Cord became chairman of the board and the other officers elected were: L.L. Young, vice president; Raymond S. Pruitt, secretary; and T.J. Dunnion, treasurer. The new board also included Lester D. Seymour, president of American Airways, the operating subsidiary of Aviation Corporation.

1933

	1933 Model 8-101	**1933 Model 12-161A**
Wheelbase (in.)	127	133
Price	$845	$1,395
No. of Cylinders / Engine	IL-8	V-12
Bore x Stroke (in.)	3.00 x 4.75	3.12 x 4.25
Horsepower	98 adv, 28.8 SAE	160 adv, 46.8 SAE
Body Styles	6	6
Other Features		

	1933 Salon 8-105	**1933 Salon 12-165**
Wheelbase (in.)	127	133
Price	$1,245	$1,745
No. of Cylinders / Engine	IL-8	V-12
Bore x Stroke (in.)	3.00 x 4.75	3.12 x 4.25
Horsepower	100 adv, 28.8 SAE	160 adv, 46.8 SAE
Body Styles	5	5
Other Features	Hydraulic internal expanding brakes against Centrifuse brake drums on four wheels, supplemented by mechanical parking brake application on the rear wheels. Trayor threaded spring shackles, produced by Tryon, lubricated by the central Bijur lubrication system. Oldberg three-tube-type mufflers. Balloon tires mounted on wood spoke wheels sizes 6.00x17 on the five-wheel cars and 6.50x17 on the six-wheel cars.	

Two new cars, the Salon 8-105 and the Salon 12-165 were added to the Auburn model lineup for 1933. The standard and custom models were continued at prices slightly above those of the 1932 models. The engines, transmissions, and rear axles (including the improved dual-ratio mechanism) were continued in the new models without material change; however, the bodies were strikingly new, with more acute angle V radiator shell, with vertical blade grilles on the 8-105 and horizontal stainless blades on the 12-165. The front A-member of the frame and spring horn cross member were effectively concealed by the unique splash apron curving down and forward of the new radiator shell, embossed in a manner resembling diagonal louvres. A new obtuse angle chromium-plated bumper bar portrayed the likeness to that of an aircraft propeller.

A stainless-steel decorative molding was carried around both front and rear fenders, outlining the edges and protecting against scratches. A similar stainless-steel molding was used on the upper half of the body sides, above and below the door and rear-quarter windows, paralleling the window lines at the windshield post (A-pillar) and back of the rear-quarter window. The cowl was entirely new with a pointed windshield header panel to accommodate the adjustable V-type windshield.

The interiors were interestingly designed, a marked advance over any previous arrangement. The round dial instruments were edge-lighted, and the panel carried the control lever for the innovative B-K power brake boost regulator; the four regulating positions were marked "Dry Weather," "Rain," "Snow," and "Ice."

In spite of a very fine product and great value offered in the 1933 Auburn model lines, the domestic sales plummeted to a dismal 5,038 units. There wasn't reason to despair because the merciless Depression and the coincident "bank holidays" during 1933 dried up most capital. Every automobile manufacturer suffered great losses in sales and revenue during 1933, with the exception of Plymouth and Dodge. Plymouth introduced a new car for 1933, with a six-cylinder engine. The 1933 Dodge was new and downsized, sharing many assemblies and components with Plymouth. Auburn had reported a six-month loss of $1,109,557 on May 31, 1933.

E.L. Cord (left) with Lucius B. Manning in 1933.

Even though the losses hindered Auburn operations, more than $1,000,000 was spent in tooling and dies for the 1934 production. Thus, the accumulated Auburn Automobile Company net loss for 1933 exceeded $2,000,000. In an austerity move, the Limousine body division was shut down, and all body operations were moved to Connersville, Indiana, as were the Auburn Automobile Company general offices. In order to consolidate resources, the Auburn plant and buildings were used to house service parts and for the manufacture of the Auburn 12-165 during 1934 using the Auburn 12-165 parts and components on hand.

1934

	1934 Standard 6-52	1934 Custom 6-52
Wheelbase (in.)	119	119
Price	$745	$845
No. of Cylinders / Engine	6	6
Bore x Stroke (in.)	3.06 x 4.75	3.06 x 4.75
Horsepower	85 adv, 22.5 SAE	85 adv, 22.5 SAE
Body Styles	3	4
Other Features	telescoping direct-acting shock absorbers, safety-glass windshield, luggage compartment, adjustable front seats and steering column	dual-ratio rear axle, telescoping direct-acting shock absorbers, safety-glass windshield, luggage compartment, adjustable front seats and steering column

1934 Auburn 8-50Y phaeton sedan.

	1934 Standard 8-50	**1934 Custom 8-50**	**1934 Salon 12-165**
Wheelbase (in.)	126	126	133
Price	$995	$1,125	$1,645
No. of Cylinders / Engine	IL-8	IL-8	V-12
Bore x Stroke (in.)	3.06 x 4.75	3.06 x 4.75	3.12 x 4.25
Horsepower	100 adv, 30.0 SAE	115 adv, 30.0 SAE	160 adv, 46.8 SAE
Body Styles	3	4	4
Other Features	telescoping direct-acting shock absorbers, safety-glass windshield, luggage compartment, adjustable front seats and steering column	dual-ratio rear axle, vacuum power brakes, telescoping direct-acting shock absorbers, safety-glass windshield, dual windshield wipers, luggage compartment, edge-lighting instrument panel, adjustable front seats and steering column	vacuum power brakes, telescoping direct-acting shock absorbers, safety-glass windshield, dual windshield wipers, luggage compartment, edge-lighting instrument panel, adjustable front seats and steering column

On January 1, 1934, the new 1934 Auburn model lines were introduced, consisting of the Auburn 6-52, 8-50, and 12-165 models. The 12-165 was a carryover model of the 1933 model with a revised sloping radiator shell and grille. The 6-52 reinstated the Lycoming six-cylinder L-head engine of 85 hp. The custom 6-52Y and 8-50Y featured entirely new design and styling of the bodies and new mechanical features and innovations. The Auburn 12-165 was continued substantially unchanged.

The body styling lines were distinctively new, influenced by the trend of the automobile industry toward streamlining (although not as severe as the "audacious" Chrysler and DeSoto Airflow and the Hupmobile Aerodyne). The radiator V-shaped fronts had die-cast grilles with rectangular openings, silver-lacquered on the standard models and chromium-plated on the Custom models and the 12-165. The hoods were somewhat longer, extending to overlap the cowl panels, producing very narrow cowl upper and side panels, without decreasing front door entrance accessibility. Windshields were one-piece safety glass, permanently installed and rubber-sealed, and taller to improve visibility, especially for traffic signal observation.

The national economy improved only slightly during 1934, thus Auburn was able to increase the car sales by just 498 units, for a total sales for 1934 of 5,538 cars. This caused some finger-pointing at Auburn, and Alan Leamy was the target and fall guy, resigning in mid-1934, but the real cause of poor sales was the economy. Fortunately, Harley Earle of General Motors admired Leamy's styling artistry, and was influential in Leamy's hiring as a designer at the Fisher Body Division. A year later, Leamy was promoted to chief stylist of the LaSalle styling unit of General Motors. Unfortunately, he died suddenly of accidental septicema from a medical injection.

Alan H. Leamy, designer for the Auburn Automobile Company from 1928 to 1934, shown here at the company's headquarters in Auburn, Indiana, in the 1931 Auburn 8-98 speedster that he designed.

Announcements

As mentioned earlier, Leamy designed the Cord L29 and the radiator shell, hood, and the sweeping front fenders of the Duesenberg J. During 1930, Gordon Buehrig joined Duesenberg Incorporated as chief stylist, and thus had a close working relationship with Leamy. They respected each other's artistic skill, so it was only natural that Gordon Buehrig would be given the task of refining the Auburn for 1935. The new 1935

Gordon M. Buehrig, designer of Auburn, Cord, and Duesenberg automobiles, shown here at the Auburn-Cord-Duesenberg Festival in 1977 with the 1933 Duesenberg SJ Arlington sedan.

Harold T. Ames, vice president of the Auburn Automobile Company, formerly with Duesenberg, Inc.

Auburn models had a pre-announcement showing for the dealers at the Connersville plant on August 24, 1934. At this showing, L.B. Manning, executive vice president of the Cord Corporation, announced the return of Roy H. Faulkner as president of the Auburn Automobile Company, causing much celebration and joy.

In April 1934 the Cord Corporation announced the formation of a new aviation operating company to supplant the American Airways. This new company was amply financed and had the personnel and equipment to make it eligible to bid for United States air mail contracts. The Cord Corporation, which owned the American Airways and the parent company Aviation Corporation, financed the new venture as a separate entity. On September 1, 1934, Lucius B. Manning was elected president of the Cord Corporation, and W.H. Beal was named vice president and assistant to Mr. Manning. In February 1935 Harold T. Ames was elected executive vice president and a member of the board of directors of the Auburn Automobile Company.

1935

	1935 Series 653	1935 Series 851
Wheelbase (in.)	120	127
Price	$745	$1,095
No. of Cylinders / Engine	6	IL-8
Bore x Stroke (in.)	3.06 x 4.75	3.06 x 4.75
Horsepower	85 adv, 22.5 SAE	115 adv, 30.0 SAE
Body Styles	5	5
Other Features	wire spoke wheels	wire spoke wheels

	1935 Salon 851	1935 Supercharged 851
Wheelbase (in.)	127	127
Price	$1,268	$1,445
No. of Cylinders / Engine	IL-8	IL-8
Bore x Stroke (in.)	3.06 x 4.75	3.06 x 4.75
Horsepower	115 adv, 30.0 SAE	150 adv, 30.0 SAE
Body Styles	5	6
Other Features	wire spoke wheels	wire spoke wheels

The 1935 Auburn models consisted of two series: the 653 and the 851, available as the standard line, custom lines and Salon lines. Although the specifications and mechanical details were unchanged, the styling was slightly modified by Gordon Buehrig. The radiator shell was more erect, the grilles were die-cast in a small rectangular pattern, the model numbers were displayed on the front of the grille, the hood side louvres were vertical with five horizontal stainless-steel moldings overlaid, the embossing was removed from the fender skirts, the fenders assumed a half-pontoon shape, the free-standing headlamps were chromium-plated and bullet-shaped, and the radiator shell was adorned with a flying girl mascot.

On January 5, 1935, at the National Automobile Show in New York, the new Auburn 851 Supercharged series was introduced, including the fashionable "Supercharged Speedster," even today one of the most sought-after classics. All Auburn supercharged 851 models were certified to be capable of 100 mph.

The unique feature of the Auburn centrifugal-type supercharger was the drive mechanism. The crankshaft to auxiliary shaft speed was increased by 1.2:1 by means of the timing chain sprocket. The auxiliary shaft was geared by 1:1 bevel gears to the vertical shaft. The vertical shaft speed was increased by a 5:1 friction-type planetary assembly, using rollers as planets instead of gears for silent operation. The planetary unit gave the sun (center) roller the impeller shaft and impeller speed the equivalent of 6:1 ratio in relation to crankshaft speed. Leading the Auburn Supercharged 851 series was the new speedster model, with four exposed, stainless-steel covered exhaust collector pipes merging into the common exhaust pipe, muffler, and tail pipe under the car.

In spite of the extensive tooling costs, the introduction of the supercharger, and the introduction of body types, Auburn domestic sales in 1935 had decreased by about 373 units to approximately 5,163 cars. The lack of sales was due primarily to the loss of dealer representation, from a peak of 1,117 dealers in 1931, to 780 in 1932, and down to a dismal 499 in 1935.

During July 1935 Ab Jenkins, famed world-record holder, drove a stock 1935 supercharged Auburn 851 over the Bonneville Salt Flats in Utah, smashing 70 unlimited and American speed records for stock cars. Jenkins averaged 104.395 mph for the first five miles, 103.69 for one hour, 103.033 mph for 500 miles, and 102.77 mph for 1,000 miles. Jenkins also broke all records from 1 to 3,000 kilometers, 1 to 2,000 miles, 12- and 24-hour records.

While 1935 was a disappointing year for the Duesenbergs at Indianapolis, they were amply rewarded by the ultimate success of the Duesenberg "Mormon Meteor" later that year. It was created through the combined efforts of August and Denny Duesenberg and J. Herbert Newport. To obtain favorable publicity the Mormon Meteor had to be a stock-chassis street-driveable vehicle. A 142-1/2-in. wheelbase Duesenberg chassis was used with SJ engine. Newport designed and styled the vehicle, while the Duesenbergs fabricated the aerodynamic creation, with pontoon-type fenders and all protruding parts fared into the coachwork or fenders.

For the record run the bumpers, headlamps, and special muffler were removed, not for aerodynamic reasons, but to avoid them falling off during the run. Driven by Ab Jenkins, mayor of Salt Lake City, the Mormon Meteor attained speeds of 160 mph during practice on the 10-mile circular course of the famed Bonneville Salt Flats, Utah. The record run was made on August 31, 1935, under the sanction and timing of the American Automobile Association Contest Board. The Mormon Meteor averaged 152.45 mph during the first hour, and 135.5 mph for the 24-hours elapsed time, including stops for fuel and tires every 400 miles. While the record was subsequently broken by a special non-production vehicle powered by a massive aircraft engine more than double the size of the Duesenberg SJ engine, it did not belittle the accomplishment of the ultimate point of greatness in Duesenberg's career. It did once more confirm the status of Duesenberg as the most successful great car built by Americans. E.L. Cord was justly proud.

In October 1935 Auburn introduced a specialty twelve-passenger limousine-type bus with graceful styling and luxurious interior.

1936

	1936 Series 654	1936 Salon 654
Wheelbase (in.)	120	120
Price	$795	$990
No. of Cylinders / Engine	6	6
Bore x Stroke (in.)	3.06 x 4.75	3.06 x 4.75
Horsepower	85 adv, 22.5 SAE	85 adv, 22.5 SAE
Body Styles	5	5
Other Features		

	1936 Standard 852	1936 Salon 852
Wheelbase (in.)	127	127
Price	$1,095	$1,268
No. of Cylinders / Engine	IL-8	IL-8
Bore x Stroke (in.)	3.06 x 4.75	3.06 x 4.75
Horsepower	115 adv, 30.0 SAE	115 adv, 30.0 SAE
Body Styles	5	5
Other Features		

	1936 Supercharged 852	1936 Cord 810
Wheelbase (in.)	127	125
Price	$1,545	$1,995-2,145
No. of Cylinders / Engine	IL-8	V-8
Bore x Stroke (in.)	3.06 x 4.75	3.50 x 3.75
Horsepower	150 adv, 30.0 SAE	125 adv, 39.2 SAE
Body Styles	5	4
Other Features		broadcloth or leather interior, adjustable front seats, no running boards, 50-in. height

The disappointing sales of Auburn during 1935 did not deter E.L. Cord's planning and innovation of the new 1936 car. The Auburn 654 and 852 introduced for 1936 were essentially carryover models of the 1935 Auburn 653 and 851. However, there was an entirely new car announced on November 4, 1935, which would again startle the automobile industry.

1936 Cord 810 phaeton.

The entirely new 1936 Cord 810, with distinctive body styling, innovative front drive, and many new features, had been groomed as the leader of the Auburn line for 1936. It represented several years of engineering development and styling, and incorporated many chassis features that were available for the first time. A top speed of more than 95 mph was claimed, and the new independent front suspension was said to provide excellent riding qualities.

The bodies were attached to a front substructure, consisting of a new system of independent suspension by means of a single transverse leaf spring in combination with two support arms, one on each side, pivoted at the front cross-member and having the yoke at the free rear end, attached to the Rzeppa constant-velocity joint at the wheel. The Rzeppa joints were held in place by the yoke ends of the arms, anchored by means of tapered pins at the knuckle. The kingpin inclination angle was such that the kingpin axis intersected the center of the tire contact with the surface of the road.

The 1936 Cord 810 was powered by a specially designed Lycoming V-8 L-head engine. The transmission had four forward speeds, incorporating constant-mesh fine helical gears. Gear control was by a remote finger-operated lever on the steering column, controlling the "vacuum-electric" power unit (located on top of the transmission) developed by Bendix Corporation.

The accessibility of the power plant and drive mechanism was a major design goal. The hood and related components consist of the hood top hinged at the cowl, the horizontally louvred hood sides integral with the front, and a transmission front cover. The hood top, surrounding sides and front, or the transmission cover were easily removed.

In spite of the innovation of the 1936 Cord 810 and the great values of the Auburn 654 and 852, the sales dwindled to only 1,174 domestic Cord sales and 1,848 domestic Auburn sales, for a total of only 3,022 units during 1936. This loss of sales caused a net loss of approximately $1,522,844 for 1936, on top of the loss of about $2,697,852 in 1935 (in which the tooling costs for the 1936 Cord 810 were included).

1936 Cord Sportsman.

1937-1939: The End of an Era

	1937 Series 654	1937 Series 852
Wheelbase (in.)	120	127
Price	$745-995	$995-1,545
No. of Cylinders / Engine	6	IL-8
Bore x Stroke (in.)	3.06 x 4.75	3.06 x 4.75
Horsepower	85 adv, 22.5 SAE	115 adv, 30.0 SAE
Body Styles	5	5
Other Features		

	1937 Cord 810	1937 Cord SC 812
Wheelbase (in.)	125	125-132
Price	$1,995-2,195	$2,195-2,795
No. of Cylinders / Engine	V-8	V-8
Bore x Stroke (in.)	3.50 x 3.75	3.50 x 3.75
Horsepower	125 adv, 39.2 SAE	170 adv, 39.2 SAE
Body Styles	4	4
Other Features		

However, on October 31, 1936, it was announced that the 1937 Cord 812 would be offered with the choice of a naturally aspirated 125-hp engine or a new supercharged engine of 170 hp. The exterior styling was unchanged, except the 170-hp Supercharged Cord 812 had two stainless-steel-covered exhaust pipes extended from each side of the hood, through the front fender inner skirt to connect with a common Y-shaped exhaust pipe under the car, connected to an intermediate pipe, to the transverse muffler under the rear body panel.

President Roy Faulkner (center), vice president Louis Jones (left), and A.H. McInnis (right), with the first 1937 Cord 812 supercharged phaeton built at the Connersville factory.

During 1937 the sales plunged even further: only 146 Auburn cars were sold domestically, and Cord accounted for only 1,147 car sales. These numbers were devastating for the Auburn Automobile Company and the Cord Corporation. While the Lycoming Manufacturing Company was engaged in the manufacture of aircraft engines and was operating profitably, the other divisions were losing money. This forced the Central Manufacturing Company to become engaged in the manufacture of automotive stampings for other automobile manufacturers, and of non-automotive metal products such as metal kitchen cabinets, refrigerators, dishwashers, and other appliances, merchandised by chain retail stores.

Considering the adverse business performance, E.L. Cord respected and accepted the suggestion of Lucius B. Manning to discontinue the manufacture of automobiles, and to sell his interest in the holdings of the Cord Corporation. On August 7, 1937, E.L. Cord agreed to sell his entire stock holdings in the Cord Corporation to the banking firms and Lucius B. Manning for $2,632,000. Automobile production stopped on that day.

The banking firms included: Schroeder, Rockefeller and Company; Victor Emmanuel and Company; Miller and George of Providence, Rhode Island. Added to the Cord Corporation board of directors were: Victor Emmanuel, Gerald E. Donovan, Thomas M. Girdler, C.C. Darling, Henry Lockhart Jr., and R.S. Pruitt. The resignations from the board were those of E.L. Cord, Harold T. Ames, L.K. Grant, and R.P. Willis.

The new board of directors met on September 15, 1937, at Cord headquarters, at which time a decision was reached to retain the profitable units of the corporation and dispose of the others. The Central Manufacturing Company division continued to manufacture production stampings, and expanded their lines of household appliances and air-conditioning units. The Lycoming Manufacturing Company continued to build and sell space heaters, and Lycoming aircraft, automobile, truck, and marine engines. The American Airlines, a unit of Aviation Corporation, enjoyed rapid growth of airline passenger travel. The Vultee division of Aviation Corporation had orders exceeding capacity for the next 20 months. The New York Shipbuilding Corporation continued to build vessels to meet the United States Navy and Merchant Marine requirements.

In February 1938 the name of Cord Corporation was changed to Aviation Corporation of America (AVCO). On July 9, 1938, Dallas Winslow purchased the Auburn Showroom and the residue of the Auburn and Cord parts and components. Glenn Pray obtained the Auburn and Cord patent rights. The north multi-story buildings in Auburn were sold to the Borg-Warner Corporation. The remaining parts for Duesenberg were purchased by Marshal Merkes, Imperial Manufacturing Company. The surplus Auburn V-12 engines were sold to American-LaFrance, fire engine manufacturers, who later ordered Lycoming V-12 engines with a larger displacement. The dies and production tooling for the Cord went to Norman DeVaux, and were later used in the manufacture of the Hupp Skylark and the Graham Hollywood models built at the Graham plant.

On November 15, 1939, all the assets of the Lycoming Manufacturing Company were sold to AVCO. During 1940 through 1945, the AVCO Central Manufacturing division was engaged in the production of war materiel. The American Kitchens division appliance business was sold to the Design and Manufacturing Company on February 8, 1960. AVCO sold the remainder of the Connersville plants and moved the ordnance manufacturing operations to Richmond, Indiana.

Through a series of mergers and acquisitions, AVCO has become the Lycoming-Textron division of Textron Incorporated, which is engaged primarily in the manufacture of military and commercial aircraft propulsion equipment. Lycoming-Textron is the foremost producer of turbojet engines for many helicopters, in addition to other aircraft engines. In fact, it was the Lycoming turbojet engine that powered the "Miss Budweiser" (unlimited) racing hydroplane to over 92 victories, 14 world championships, and 9 gold trophies.

As for E.L. Cord, he must have taken the advice of Horace Greeley to "Go west, young man." He moved to Southern California and took up a new enterprise — real estate. He also owned an all-music radio station. After World War II, he was successfully engaged in uranium mining in Utah. During the 1950s, he was actively involved in politics and was elected a state senator in Nevada. He was later heavily involved in the election of John F. Kennedy as President of the United States. E.L. Cord passed away at his home in California in 1974 at the age of 80.

E.L. Cord (left) served as state senator in Nevada in 1956 as a Democrat. He is shown here on a plane trip to Texas with John F. Kennedy (right) and Hubert Humphrey (center).

Index

Page numbers followed by *p* indicate a photo.

AC-Delco division, GM, 106
Acetylene gas products, 102
Aero-Willys, 84, 85*p*
Air transportation
    American Airlines, 189
    Aviation Corporation, 179–180, 184
    Century Air Lines, 177
    Eastern Airlines, 157
    Stinson Aircraft Corporation, 172
Aircraft engines
    Allison, 103
    King-Bugatti, 68
    Liberty, 44–46, 103
    Sunbeam, 64
Aitken, John, 105
Alden Sampson Manufacturing Company, 91
Allison Engineering Company, 103
Allison, James A., 101*p*, 102*p*, 157*p*
    background, 101, 102, 103
    Empire Motor and, 105
    Indianapolis Motor Speedway and, 106, 112
American Airlines, 189
American Bicycle Company, 21, 24, 104
American Motor Car Sales Company, 52–53
American Motors Corporation, 84–85
Americar by Willys-Overland, 83*p*
Ames, Harold T., 179, 184, 184*p*
Argo Motor Company, 93
Army, U.S.
    contract with Lincoln, 44–46
    purchase of White trucks, 12–13, 14
Atlantic City Speedway, 96
Atwell, W.S., 20
Auburn Automobile Company
    acquisitions, 170, 172
    business status during Depression, 174, 176, 179
    Cord's arrival, 167, 169
    Cord's product line changes, 168–169
    decline in sales (1937-39), 188–189
    8-98 debut, 176–177
    front-drive passenger car, 173*p*, 173–174
    ownership change, 166
    personnel changes, 177–178
    product line and prices (1903-17), 161–164*p*, 161–166
    product line and prices (1920-24), 166–167, 167*p*, 168*p*
    product line and prices (1925-27), 169
    product line and prices (1927-29), 170–171, 171*p*
    product line and prices (1930-31), 175, 175*p*, 176*p*
    product line and prices (1932), 178
    product line and prices (1933-34), 180–182, 182*p*, 183*p*
    product line and prices (1935), 184–185

    product line and prices (1936), 186*p*, 186–187, 187*p*
    product line and prices (1937-39), 188*p*, 188–189
    start of business, 160–161
Auto races
    Auburn car speeds, 171
    Devil's Despair Hill Climb, 96
    Indianapolis Motor Speedway start, 102
    "Mormon Meteor" record, 185
    National cars at Indianapolis, 105
    New York to Paris, 126
    Pope-Hartford models, 31
    White Steamer participation, 5–6
Auto shows
    National Automobile Show. *See* National Automobile
      Show
    George Pope's involvement in, 33
Autocar Company, 17
Avery, P.C., 102
Aviation Corporation, 179–180, 184
Aviation Corporation of American (AVCO), 189

"Baby Overland," 64, 67*p*
Baldwin, Henry, 25*p*
Barthel, Oliver E., 39–40
Baster, F.S., 13
Bauman, J.H., 13
Beal, W. Huber, 177, 179, 184
Bean, Ashton G., 11, 12, 13
"Beauty Six," 166, 167*p*
Bender, F.M., 13
Bicycles
    Indiana Bicycle, 21
    Lozier Company, 141
    Newby production, 104
    Pope production, 20
    White production, 4
    Willys' business, 52
Big Sandy Coal and Coke Company, 133
Black, Clarence A., 37, 38, 42
Black, Robert F., 13, 13*p*
"Blue Streak," 100
Bourquin, James F., 135
Bowen, Lem W., 38, 42
Bramwell-Robinson Sociable, 25*p*, 25–26
Bramwell, W.C., 26
Breer, Carl, 71
Briscoe, Benjamin, 87*p*
    business background, 87
    entry into auto business, 89

Briscoe, Benjamin *(cont.)*
    exit from company, 94
    expansion attempt, 89–90
    new ventures (1913-15), 93–94
    United States Motor founding, 91–92
Briscoe, Frank
    business background, 87, 89, 91, 93
    new ventures (1913-15), 93–94
Briscoe Freres, 93
Briscoe Manufacturing Company, 87, 88*p*
Briscoe Motor Company, 94
Bromley, W.K., 117
Brown, David Bruce, 105
"Brunn" by Lincoln, 48*p*
Brush, Alanson P., 39, 40, 89
Brush Runabout Company, 89, 90*p*, 91
Buehrig, Gordon, 183, 183*p*
Buick, David Dunbar, 89
Buick Motor Company, 42
Buses by White, 12, 12*p*

Cadillac Automobile Company
    closed car adoption, 42*p*, 43
    Leland association, 39
    merger with General Motors, 42
    merger with Leland, 40
    plant fire, 40, 41
    product line and prices (1903-09), 40
    start after Detroit Auto, 38
Campbell, E.B., 52, 54
Campbell, Henry, 58, 59
Canada and White Company, 8
Canaday, Ward M., 64, 80, 82
Carburetors, 106
Central Manufacturing Company, 172, 189
Century Air Lines Incorporated, 177
Chalkis Manufacturing Company, 131
Chalmers-Detroit Company
    car photos, 128–130*p*, 131*p*
    military business, 131
    product line and prices, 132
    reorganization under Maxwell, 130–131
    start of company, 127
Chalmers, Hugh
    background, 125*p*, 125
    death, 131
    exit from Chalmers-Detroit, 130
    interest in Thomas-Detroit, 126–127
Chandler-Cleveland Motors Corporation
    product line and prices, 148–150, 150*p*
    sale to Hupp, 150
Chandler, Frederick, 141*p*
    business background, 141–142
    car company start, 143
Chandler, Harry, 156
Chandler Motor Car Company, 144*p*
    Cleveland Auto association, 144
    incorporation and officers, 143

    product line and prices (1914-17), 143
    product line and prices (1919-26), 144–148, 147*p*
    product line and prices (1927-29), 148–150
    sale to Hupp, 150
    technical advances, 143
Chapin, Roy Dikemam
    background, 125–126, 126*p*
    interest in Thomas-Detroit, 126–127
Chapman, Fred S., 25*p*
Chapman, John W., 25*p*
Charles A. Strelinger and Company, 38–39
Chase, W.W., 8
Chrysler Corporation, 85
Chrysler Six, 71
Chrysler, Walter P.
    move to Chalmers, 131
    move to Willys-Overland, 68–69, 70
Church, H.D., 11
Cleveland Automobile Company
    incorporation and officers, 144
    product line and prices, 144–148, 147*p*
    products after Chandler merger, 148–150
Closed automobile bodies, 42*p*, 43
Coffin, Howard E., 126*p*
    background, 126
    interest in Thomas-Detroit, 126–127
Cohu, LaMotte T., 179
Colonial Laundry Company, 140
Columbia Automobile Company, 25, 26
Columbia bicycle, 20*p*
Columbia electric car, 23, 24*p*
Columbia Mark LXVI, 28–29
Columbia Motor Car Company
    incorporation and officers, 30
    merger with Maxwell, 91
Concentrated Acetylene Company, 102
Cooley, William B., 30
Cord Corporation
    incorporation and officers, 172–173
    new aviation operating company, 184
    sale of assets (1937), 189
Cord, E.L., 159*p*, 181*p*
    campaign for Aviation control, 179–180
    facilities purchase, 172
    move to Auburn, 167–169
    post-auto career, 190, 190*p*
    sale of company, 189
Cord L29, 174
Cost of vehicles
    Auburn cars. *See* Auburn Automobile Company, product
        line and prices
    bicycles in 1877, 20
    Briscoe cars, 94
    Cadillacs, 40
    Chalmers cars, 132
    Chandler and Cleveland cars, 144–148
    Chandler cars, 143
    Edwards-Knight car, 62
    Empire Motor car, 105
    Falcon-Knight Six, 75

Garford 60 Town Car, 59
Matheson cars, 96, 98, 99
Overland cars, 64, 70, 73
Paige-Detroit, 135, 136p, 137, 138
Parry cars, 117
Pathfinder cars, 119, 120
Pope-Hartford models, 31, 33
Pope-Robinson cars, 26, 28
Salon Limousine, 43
White cars, 8
Willys cars, 76, 79
see also individual companies
Cox, Claude E., 52, 54–55, 115–116
Cugnot, Nicholas, 1
Cunningham, Harry, 153
Curtiss OX-5, 64
Cyclops headlamp, 94

Dahl, T.R., 11, 13
Dawson, Joe, 105, 111, 112p
Day, George H., 23
Dayton Motor Car Company, 92
DeBoutteville, DeLamarre, 2
DePalma, Ralph, 111, 112p
Depression's impact on auto companies
    Auburn, 174, 176, 179
    Duesenberg Motors, 174
    Indianapolis Motor Speedway, 112–113
    Lincoln, 46–47
    Overland, 54–55
    Paige-Detroit, 137
    White, 12, 13
    Willys, 68, 76–78, 79
DeRivaz, Isaac, 1
DeSoto Motor Corporation, 100
Detroit Automobile Company, 37–38
Devil's Despair Hill Climb, 96
DeWar Trophy, 42
Diamond Chain Company, 104
Diamond-T-Motor Truck Company, 17
Disbrow, Louis, 31, 32p
Dissolved-acetylene industry, 102
Dorn, F.R., 117
Douglas, Donald, 156
Drive mechanisms improvements, 185
Duesenberg, Fred, 155p, 171, 179
Duesenberg Motors Corporation
    acquisition by Auburn, 170
    asset purchase by Willys, 68
    business status during Depression, 174
    "Mormon Meteor," 185
Dunnion, T.J., 180
Durant, Cliff, 153
Durant Motors Incorporated, 71
Durant, William C.
    Cadillac association, 42, 43
    merger plans with Briscoe, 89–91
Duryea, Charles and Frank, 2

E-M-F Company, 38, 59
Earl, Clarence A., 63, 65, 69, 94
Earl Motors Incorporated, 94
Eastern Airlines, 157
Eckhart Carriage Company, 160, 163, 166
Eckhart, Charles, 159–160, 160p, 165
Eckhart, Morris and Frank, 160p, 166
Edenburn, Eddie, 158p
Edward G. Gudd Manufacturing Company, 59
Edwards, H.J., 59, 93
Edwards-Knight car, 60–62
Edwards Motor Car Company, 60
Electric Auto-Lite Corporation, 71
Electric Carriage and Wagon Company, 21–22
Electric cars
    Columbia Mark LXVI, 28–29
    industry consolidation, 23
    by National, 104
    by Pope, 23, 24p
    popularity of, 21
    by Riker, 23, 26
Electric Storage Battery Company (Exide), 22, 26
Electric Vehicle Company
    background, 22–23, 25p
    Columbia Mark LXVI, 28–29
    reorganization (1909), 30
Electro Metallurgical Company, 102
Electrobat, 21, 22p
Elmira Arms Company, 52
Emergency Banking Law, 79
Emise, Charles A., 142
Empire Motor Car Company, 105
E.R. Thomas Motor Company, 93, 126
Evans, E.R., 153
Evans, Oliver, 1
Everitt-Metzger-Flanders Company (E-M-F Company), 38, 59
Everitte, Byron P. (Barney), 153
Exide (Electric Storage Battery Company), 22, 26

Falcon-Knight Corporation, 75
Falcon-Knight Six, 75, 76
Farley, J.I., 166
Faulconer, Robert C., 36, 36p
Faulkner, Roy, 166, 177p, 177–178, 188p
F.B. Stearns Company, 74, 75
Fellers, Morse, 8
Fisher, Carl Graham, 101p, 103p, 108p
    background, 101–102
    Empire Motor and, 105
    Indianapolis Motor Speedway and, 106
    Miami Beach development, 103–104
    resignation from Indianapolis, 112
Fisher Closed Body Company, 43
Flagler, Henry M., 104
Flanders, Walter, 130, 153, 154
Fletcher, Edward O., 104
Forbes, Thomas P.C., 55, 58
Ford, Edsel, 50p

Ford, Harry W., 127, 130
Ford, Henry, 50*p*
    Detroit Auto association, 37, 38, 39
    Lincoln purchase, 48
    racing interest, 40
Four-wheel brakes introduction, 154
Fox, Frank, 31, 32*p*
Frayer, Lee, 151
Frazer, Joseph W., 81, 82*p*, 82
Front-drive passenger car, 173*p*, 173–174

G-A touring car by White, 9*p*
G-B Landaulet by White, 9*p*
Garford 60 Town Car, 59
Garford, Arthur L., 59, 59*p*
Garford Company, 59, 60*p*
General Motors Company
    AC-Delco division, 106
    incorporation and officers, 42, 91
    purchase of Allison, 103
General Purpose vehicle (GP, Jeep), 82, 82*p*, 83*p*
George, Mari Hulman, 113
George, Tony, 113
Gilcrest, E.H., 166
Globe Ball Bearing Company, 62
"Good Roads Movement," 20
Gorman, Martin, 41
Gormully and Jeffrey Manufacturing Company, 21
Graham-Paige Motors Corporation, 100, 140
Gramm Motor Truck Company, 59
Greuter, Charles R., 95, 96
Grout, William L., 3–4

H.A. Lozier and Company, 141–142
Half-Trac personnel carrier, 14
Hall, Elbert J., 45
Harroun, Ray, 111*p*, 111
Hassler, Robert, 105
Haynes, Elwood, 12
Henry Ford Company, 38
Henry Kaiser Corporation, 84
"Highway" Twelve by National, 104
Historical overview of auto industry, 1–2
Howe, E.C., 104
Hulet, E.W., 10
Hull, Ellen, 35
Hulman, Anton (Tony), 113, 158
Humphrey, Hubert, 190*p*
Hupp Motor Car Company, 150
Hussey, Pat, 40
Huygens, Christian, 1

Indiana Bicycle Company, 21
Indiana Motors Corporation, 12

Indianapolis Motor Speedway, 152*p*
    500 races (1911, 1912), 110–112
    inaugural races, 108
    incorporation and officers, 106
    legacy of, 113
    National cars in competition, 105
    ownership changes, 112–113, 156, 158
    plans and dimensions, 106–108, 107*p*
    resurfacing, 109, 109*p*
    start of, 102–103
Indianapolis Speedway Team Company, 102–103
International Machinists Union, 66
Inyo Chemical Company, 49

Jackson Machine Parts Company, 94
J.B.B. Colleries Company, 134
Jeep-Eagle Division, Chrysler Corporation, 85
Jeeps, 82*p*, 82, 83*p*
Jeepster, 84*p*, 84
Jefkins, Rupert, 112*p*
Jenkins, Ab, 185
Jewett, Bigelow and Brooks Company, 133, 134
Jewett, Harry Mulford, 133*p*
    background, 133
    contribution to Paige, 137
    exit from company, 140
    investment in Paige, 134
Jewett Motors Incorporated, 137
"Jo-Blocks," 42
Johansson, Carl Edward, 42
Johnson, M.B., 10
Johnson, Wilbur C., 29, 30
Jones, James J., 25*p*
Jones, Louis, 188*p*
Jones, Saunders, 11

Kaiser-Willys Corporation, 84
Kelly, George H., 11, 13
Kelvinator Corporation, 100
Kemp, A.P., 166
Kenan, L.D., 96
Kennedy, John F., 190*p*
Keyes, J.H., 116
Kilpatrick, W.R., 69
Kincaid, Thomas, 105
King-Bugatti, 68
Klein, Seth, 158*p*
Knudsen, Semon E., 17, 17*p*

Labor unrest and strike at Willys-Overland company, 66–68
Lallement, Pierre, 20, 21
Laminated safety windshield glass, 156
Law, Fred A., 25*p*
Lawrence, B.F., 108*p*

Leamy, Alan, 172, 173, 176, 183, 183*p*
"Leather Stocking" model, 118
Leland and Faulconer Manufacturing
    business start, 36–37, 37*p*
    Cadillac business dealings, 38–39
    merger with Cadillac, 40
Leland, Edith Miriam, 35, 39*p*
Leland, Gertrude, 35
Leland, Henry Martyn
    background, 35–36
    Lincoln Motor start, 44
    photos, 35*p*, 36*p*, 37*p*, 38*p*, 39*p*, 43*p*, 45*p*, 50*p*
    position at General Motors, 42
    precision engineering concept, 42
    resignation from Cadillac, 43
    resignation from Lincoln, 49
    tool business with Faulconer, 36–37
    work at Cadillac, 40, 41
Leland, Wilfred C., Jr., 38*p*, 39*p*
Leland, Wilfred Chester
    background, 35, 36, 42, 43
    Lincoln Motor start, 44
    photos, 39*p*, 43*p*, 46*p*, 50*p*
    resignation from Lincoln, 49
Lenoir, Etienne, 2
Lewis Spring and Axle Company, 94
Liberty aircraft engines, 44*p*, 44–46, 45*p*, 103
Limousine Body Company, 170
Lincoln Motor Company
    car photos, 47*p*, 48*p*, 49*p*
    financial impact of Depression, 46–47
    government contract dispute, 49
    Liberty contract with Army, 44–46
    organization and officers, 44
    purchase by Ford, 48
    V-8 engine, 46
Linde Air Products Company, 102
Linder, Abbey, 20
Lloyd, Herbert, 30
Lozier, H.A., Jr., 141, 142*p*
Lozier Motor Company, 141–142, 142*p*
Lycoming Manufacturing Company, 170, 171, 174, 176, 177, 189

M-16 truck, 15
Macauley, Alvan, 45*p*
Macrea, F.T., Jr., 13
Manning, Lucius B., 178, 180, 181*p*, 184, 189
Marcus, Siegried, 2
Marion-Handley car, 56
Marion Motor Car Company, 55
Martin-Parry Corporation, 122
Marvel carburetor, 106
Matheson, Charles W., 95*p*
    business background, 95
    post-Matheson positions, 100
Matheson Motor Car Company
    financial trouble, 98–99

    product line and prices, 96, 98, 98–99, 99*p*
    start of company, 95
Maxim, Hiram Percy, 23, 23*p*, 25*p*
Maxwell AA runabout, 91*p*
Maxwell-Briscoe Motor Company
    business start, 89
    car photos, 91*p*
    merger with Columbia, 91
Maxwell, Jonathon D., 89, 89*p*, 93
Maxwell Motor Company Incorporated, 130–131
McInnis, A.H., 188*p*
Mead, W.S.M., 142
Meeker, Ezra, 120, 121*p*
Merkes, Marshal, 189
Merz, Charles, 105
Metzger, William E., 38
Meyers, T.E., 155*p*, 158*p*
Military use of vehicles
    Jeeps, 82*p*, 82, 83*p*
    White trucks, 8, 9, 10, 12–13, 14
Miller, Linwood A., 76
Miller, Otto, 10
Miniger, Clement O., 71
Moline Plow Company, 71
Mooney, James D., 84
"Mormon Meteor," 185
Morris, Henry G., 21
Motor Car Manufacturing Company, 117–119
Mulford, Ralph, 111, 148
Murphy, William A., 37, 42
Mutual Motors Company, 56
Myers, Charlie, 148
Myers, Theodore E., 103

National Automobile and Electric Company, 104–105
National Automobile Show
    Auburn cars, 171
    Auburn Supercharged debut, 185
    Cadillac debut, 38
    8-98 debut, 176
    Rickenbaker Six, 153
National Bank Holiday, 79
National Carbon Company Incorporated, 102
National Vehicle Company, 104
"New Day" Jewett, 138–139, 139*p*
New Duesenberg J, 172
New York Motor Sales Company, 56
New York to Paris race, 126
Newby, Arthur C., 101*p*, 104*p*
    auto business, 104–105
    background, 104
    Empire Motor, 105
    Indianapolis Motor Speedway and, 102, 103, 106
Nichols, John A., 75
"999" race car, 40
Norton, Charles H., 37
Norton Company, 37
Nuckols, Henry W., 30

Oakland Motor Car Company, 42
Olds Motor Works, 42
Oliver Corporation, 17
Overland Automobile Company
　　car photos, 53p, 55p, 56p, 57p, 58p, 63p, 65p
　　financial rescue by Willys, 54–55
　　incorporation and officers, 52
　　interaction with Parry, 115–116
　　Overland Four, 72
　　Overland Whippet, 74–75, 75p
　　reorganization, 116
　　sales growth, 63–64

Paige-Detroit Motor Car Company
　　decline in sales (1926-27), 138–139
　　product line and prices, 134–135, 135p, 136p, 137–138
　　reorganization under Jewett, 134
　　sale to Graham, 140
Paige, Fred, 134, 134p
Parry Automobile Company
　　incorporation, 116
　　models and prices, 117, 117p
Parry, David M.
　　background, 115
　　death, 122
　　interest in Overland, 52, 54, 115–116
　　own auto company start, 116–117
Parry Manufacturing Company, 122
Pathfinder Automobile Company
　　incorporation and officers, 119
　　product line and prices, 119, 119p, 120p, 120, 121p
　　sale of assets, 122, 123p
Pelletier, E. LeRoy, 153
Phaeton, Lincoln, 49p
"Pike's Peak Motor," 145
Pope, Albert Augustus
　　background, 19p, 19–20
　　death, 30
　　road improvements activism, 20
Pope, Albert L., 29
Pope, Edward W., 26
Pope, George, 33, 33p
Pope Manufacturing Company
　　bicycle production and patents, 20, 21
　　business expansion, 25
　　electric cars, 23, 24p
　　end of company, 33–34
　　financial status and officers (1912), 32–33
　　incorporation, 24–25
　　Pope-Hartford models (1910-13), 31p, 31–32, 32p, 33
　　product line and prices (1903-07), 26p, 26–29, 27p, 28p
　　reorganization (1907-08), 29–30
　　sale of assets, 30, 56–57
Pope-Robinson Company, 26
Powerboat racing, 103
Pratt and Whitney Company, 33

Pray, Glenn, 189
Precision manufacturing, 42
Prest-O-Lite Company, 102
Prices of cars. *See* Cost of vehicles
Pruitt, Raymond S., 180

Regar, Samuel, 142
Reliance Motor Truck Company, 134
Reo Motors Incorporated, 17
Rice, Herbert H., 29, 30
Rice, Isaac L., 21
Rickenbaker, Edward V.
　　aircraft industry work, 156–157
　　car company start, 153
　　early racing days, 151, 152p
　　photos, 151p, 155p, 157p, 158p
　　post-auto career, 157–158
　　Speedway transactions, 103, 112–113, 156, 158
　　war years, 152
Rickenbaker Motor Company
　　business problems, 154, 156
　　four-wheel brakes introduction, 154
　　incorporation and officers, 153
　　product line and prices, 153–154, 154p, 155p, 156p
Ricker, Chester, 155p
Riker, A.L., 23
Riker Electric Vehicle Company, 23, 26
"Road Runner, " Lincoln, 47p
Robinson, John T., 26
Robinson Motor Vehicle Company, 25
Roos, D.G., 81
Royal Automobile Club, 42
Russell, George F., 13
Ryan, Ellis, 172

Salom, Pedro G., 21
Salon Limousine, 43
Sanderson, F.M., 8, 9
Saxon Motor Car Company, 127, 130
Schebler, George, 106
Schoonermobile, 120, 121p
Schultz, Glen, 148
Schwab, Charles W., 156, 157p
Scout car by White, 10p, 13, 14, 16p
Searles, W.S., 13
Selden patent, 23, 91
Self-propelled land vehicles, 1–2
Sewing machine businesses, 4, 20
Seymour, Lester D., 180
Sheridan car, 153
Simons, G.O., 117
Skelton, Owen, 71
Smith, George W., Jr., 11
Sorensen, Charles E., 82

Standard Motor Company, 93
Standard Wheel Company, 115–116
Steam-powered cars/trucks
	production phase out by White, 8
	White Steamer, 4*p*, 4–5, 7*p*
Steam tractors, 1
Stearns-Knight, 74, 75
Stellite, 12
Sterling Bicycle Company, 21
Sterling Motor Truck Corporation, 17
Stetson, Frederick L., 90
Stinson Aircraft Corporation, 172
Stoddard, Charles G., 59, 93
Stoddard-Dayton, 92, 92*p*
"Straightaway Eight," 139
Stuart, James A., 108*p*
Studebaker Corporation, 12
Studebaker-Garford 40, 59
Stutz Motor Car Company, 59
Sunbeam aircraft engines, 64
Super Power engines by White, 13, 14*p*
Sweet, Ernest E., 36
Sweetson, D.E., 25*p*

Tambly, Egbert J., 29
Teagle, Walter C., 11
Teasdale, W.C., 117
Technical advances in autos
	acetylene headlights, 102
	by Chandler, 143, 146
	closed body cars, 42*p*, 43
	drive mechanisms, 185
	four-wheel brakes introduction, 154
	front-drive passenger car, 173*p*, 173–174
	laminated safety windshield glass, 156
	precision engineering concept, 42
	by White Motor Company, 12, 13, 14*p*, 17
	by Willys-Overland Company, 78–79
Textron Incorporated, 189
Tharpe, Virginia Kirk, 177
Thermo-syphon cooling system, 89
Thomas-Detroit Company, 126, 127*p*
Thomas, E.R., 126*p*, 126
	see also E.R. Thomas Motor Company
Thomas Flyer, 126
Tidlund, Ed, 25*p*
Toledo Bicycle Company, 21
"Traffic Transmission," 146
Transmissions
	by Chandler, 146
	Paige-Detroit cars, 139
Trevithick, Richard, 1
Truck production
	Gramm, 59, 63
	Martin-Parry, 122
	White, 8, 15*p*

Union Carbide and Carbon Company, 102
United States Motor Company
	financial trouble, 92–93
	founding by Briscoe, 91–92
U.S. Post Office Department, 157

V-1710 Allison engine, 103
Vanderbeck, Herbert, 25*p*
Vertical "8," 154, 155*p*
Vincent, Jesse G., 45
Vivian, André, 1
Volvo-GM Heavy Truck Corporation, 17
VonHake, Carl, 30

Wagner, Fred, 112*p*
Walden, Herbert, 25*p*
Walker, John Brisben, 89
Warner, A.R., 8, 9, 10
"Warner Hy-Flex," 139
Watt, James, 1
Waverly Company, 30
Waverly electric car, 22*p*, 29*p*
Waverly limousine, 23*p*
Weatherhead Company, 150
Webb, Jay, 6*p*
Weed Sewing Machine Company, 20
Weidley, Walter, 120
W.G. Wilson Company, 4
Wheeler, Frank, 101*p*, 106*p*
	Indianapolis Motor Speedway and, 102, 105–106
	retirement, 112
Wheeler-Schebler Carburetor Company, 106
Wheeler, W.A., 139
Whippet by Overland, 74–75, 75*p*
Whistling Willy, 6*p*, 6
Whitbeck, John V., 142
White, Albert E., 37
White Company
	expansion to Canada, 8
	organization and officers, 8
	product line, 8, 9*p*
	production (1906-08, 1912), 8
	production (1913-14), 9
	steam engine phase out, 8
	see also White Motor Company
White Consolidated Industries, 17
White Farm Equipment Company, 17
White Manufacturing Company, 3–4
White Motor Company
	business purchases, 17
	financial impact of Depression, 12, 13
	incorporation and officers, 10, 13
	military business, 8, 9, 10, 12–13, 14, 16*p*
	production (1917-25), 11

White Motor Company *(cont.)*
    production (1926-30), 12
    production (1935, 1937), 13
    reorganization (1935), 13
    technical advances, 12, 13, 14p, 17
    truck and tractor lines dominance, 17
White, Rollin H., 4, 9, 17
White Sewing Machine Company
    auto company spinoff, 8
    expansion to Europe, 6
    military business, 7p
    move into auto manufacturing, 4–5, 5p
    start of, 4
White Steamers, 4, 4p, 7p
    auto races, 5–6
    production (1906-08), 8
White, Thomas H., 3–4, 9
White, Thomas II, 11
White, Walter C., 5, 8, 9, 10–11, 11p
White, Windsor T., 4, 8, 9, 10, 11, 17
WhiteHorse delivery vehicle, 13
Whitney, William C., 22
Wilcox, Howdy, 105
Willis, Paul, 108p
Willys, John N., 51p, 56p
    ambassadorship, 78
    background, 51–52
    car sales entry, 52–54
    death, 80
    management style, 62–63
    Marion Motor purchase, 55
    Overland rescue, 54–55, 116
    Pope-Toledo plant purchase, 30, 56–57
    retirement, 76
    return to company, 78
Willys-Knight cars
    66D, 78p
    models and prices, 62, 62p, 70p, 72, 73p, 74p, 77p, 78p
    sales growth, 63–64
    Standard Six, 76
Willys Motors Incorporated. *See* Willys-Overland Company
Willys-Overland Company
    acquisitions and contracts, 59–62, 63, 71, 74
    car photos, 57p, 58p, 63p, 65p, 67p, 78p, 80p, 81p, 82–85p

Chrysler Six plans, 71
Chrysler's changes to, 69
Duesenberg purchase, 68
financial impact of Depression, 68, 69, 76–78, 79
financial recovery (1921), 70
financial status (1913), 62
Garford cars, 59, 60p
incorporation and officers, 57
Jeeps, 82p, 82, 83p
Knight cars. *See* Willys-Knight cars
labor unrest and strike, 66–68
mergers (1950s), 84–85
military business, 64–65, 68, 82, 82–83p
new models (1936-39), 80–82, 81p
prices of cars, 59, 64, 73
product line and production (1920s), 72p, 72–76, 73p, 74p, 75p
production (1909-12), 55, 57–58p, 57–59
production (1915), 63
profit-sharing plan, 65
receivership (1933), 79–80
Red and Blue birds, 72p, 73p
reorganization (1921-23), 70–72
reorganization (1939), 81
technical advances (1932), 78–79
Willys' retirement, 76
see also Overland Automobile Company
Wilson, Charles B., 69, 71
Wilson, David R., 80, 81p
Winslow, Dallas, 189
Woodruff, Robert W., 11
World War I and White truck production, 10

York Motor Express Corporation, 122
York Trailer Company Limited, 122
Young, L.L., 180

Zeder, Fred, 71

About the Author

Mr. Michael J. Kollins was born on March 20, 1912, in St. Clairsville, Ohio, the sixth of seven children of Michael A. and Marian (Peck) Kollins. His father held franchises in that rural area of southeast Ohio to sell and service Buick automobiles in 1910, and later several other makes over several decades. Michael Kollins learned of automobile manufacturers early, literally from reading hubcap designations in his father's shop. Chores included cleaning automotive parts, scraping and installing bearings and grind valves as necessary. He also ordered, stocked, and sold parts. His father promoted wide reading of industry magazines of the early period. Michael gained "on the job" training in the automotive industry, intensely aware of many service capabilities given design and manufacturing concerns over quality matters.

Michael and Julia Kollins.

Mr. Kollins has had a unique and long employment history with the industry. As economic times were difficult in his local area, he moved to bustling Detroit in his late teens to join an older brother. He soon became a "co-op" engineering student in the late 1920s at the College of the City of Detroit, now Wayne State University. He began work as a service technician at the Dodge Brothers, then, still a "co-op" student, moved to the Packard Motor Car Company in 1930 as a test driver and engineering analyst of Packard and competitive models, a position that gave him knowledge of most of the cars manufactured in America. He continued at Packard after graduation in 1932, rising to be Chief of Section, Technical Data and Service Engineering from 1945 to 1955. With the demise of Packard, he joined Chrysler in similar activity: He was Service Technical Manager (1955), Manager of Warranty Administration (1964), and Manager of the Highland Park Service center (1968 to retirement in 1975).

Soon after undertaking formal engineering studies, Mr. Kollins became a local car racer on weekends. That experience, in brief, evolved into a long association with the Indianapolis Motor Speedway. Of note, he became Technical Inspector, Vice Chairman of the Technical Committee, Chairman of the Product Certification Committee, and later, Honorary Life Member of the Indianapolis 500 Oldtimers Club in 1952.

In World War II, Michael Kollins served in the U.S. Navy in the Motor Torpedo Boats Squadron 4. He was a Chief Machinists Mate, and later Ensign with engineering assignments. After the war, he continued in the Naval Reserve, retiring in 1972 as Lieutenant Commander.

For almost a half-century, Mr. Kollins has been a member of the Society of Automotive Engineers and the Engineering Society of Detroit. In the early 1980s he became a trustee with the Detroit Public Library National Automotive Collection, writing articles for its quarterly newsletter, *Wheels*. He wrote technical publications for the Navy, Packard, and Chrysler. In particular, he was co-author of a centennial publication of the Engineering Society of Detroit, *ESD Technology Century* (1998). He also is a historian of many automotive developments in the Detroit area.

Mr. Kollins married Julia Advent in 1934 and has two sons and a daughter and three grandchildren. Noticing his intense research in the late 1980s, his older son encouraged him to write this book on the engineering history of the automotive industry, pointing out that he was the logical person to record such happenings through his personal contacts and assocations over the years with several pioneers, executives, and suppliers.